LOW-IMPACT FORESTRY

LOW-IMPACT FORESTRY

Forestry as if the Future Mattered

Edited by Mitch Lansky

2002
Maine Environmental Policy Institute
Hallowell, Maine

Maine Environmental Policy Institute
220 Water Street
POB 347
Hallowell, Maine 04347
www.meepi.org
www. lowimpactforestry.org

Printed in the United States of America
Library of Congress Catalog Card Number: 2002105881
ISBN 0-9719962-0-2

Printed by: Recycled Paper Printing, Inc.
12 Channel Street, Suite 603
Boston, MA 02210
Printed on 100% postconsumer recycled paper bleached without chlorine.

Cover design by Christopher Kuntze

Front cover: “Pulling cable.” Back cover: “Limbing.” Both photos by Mitch Lansky.

Table of Contents

Acknowledgements

I thank the Hancock County Planning Commission for its support of this writing project, and for the Maine Environmental Policy Institute for agreeing to publish the book. I especially thank Ron Poitras, Sam Brown, and Geoff Zentz for their input and review, and Jim Fisher for his chapter on Coop options. For more editorial insight, I thank Don Mansius, Will Sugg, and Jean English. Kit Kuntze gets kudos for his excellent work on cover design. I thank the woodlot owners and loggers of Hancock County who participated in the early forums on Low-Impact Forestry (LIF) and regret that the Low-Impact Forestry Project was not able to serve them better. This book could not have been written in its current form without the help and input of the Northern Appalachian Restotration Project and the *Northern Forest Forum*, which published some of these chapters as articles and helped support the Low-Impact Forestry Project. It also could not have been published without the support of Will Sugg and the Maine Environmental Policy Institute. I appreciate the time Bob Matthews spent showing me how to do low-impact logging in Baxter State Park (his picture is on the cover). I am grateful for use of writing by Wendell Berry, William Ostrofsky, and Barbara Alexander, and the cooperation of David Perry, Jimmy Potter, Gordon Mott, Brooks Mills, and Mel Ames who are profiled or interviewed in the book. I extend gratitude to the hundreds of people who helped our family after our house burned in 2000, and my love to my wife Susan, who has shared her life with me for more than a quarter of a century.

This book is dedicated to Solomon Dean Hearn, my new grandson. The future does matter.

Foreword

By Ron Poitras, Hancock County Planning Commission

The tally of environmental and social costs that must be charged against industrial forestry is increasingly well known, and every day, it seems, the magnitude grows. This book goes far beyond that tally into an analysis that is sobering and at the same time inspiring. If we view the forest solely as property, then "rational resource mangement" is a logical goal to pursue. If however the future does matter, then we will need to learn a better way of fitting in with how the forest has worked for millions of years. With this volume Mitch Lansky has brought together real solutions that go beyond the realm of "rational resource mangement" to a necessary change in how we see our place in nature.

Many of us recognize the need for a better marriage between ecology and economics in our forests. Much of this book is about forest sustainability in the context of local communities where nature, not global economics, is the measure. This work resonates with Mitch's first hand experience, on-the-ground observations, and tangible research, not just the results of other people's work. It is clear that the principal author here has taken the time to really *see* our forests. Too many of us are operating solely from concepts of what is considered scientifically correct, and not enough from what our eyes tell us directly.

I have a neighbor who last year had his fifty-acre woodlot cut. This is an intelligent man, a musician whose artistic sensibility apparently does not extend to his land. Despite all the evidence before his eyes--of ground deeply rutted and residual trees badly scarred--he remains firmly convinced that he's obtained a scientifically prescribed, economically efficient harvest--but what a disaster! As educators about the forest, we have much work ahead of us. We must learn to question our blind faith in experts and remove the blinders from too much abstraction.

A lot of the material contained here deals with the difficult work of loggers. To care for the woods as these people do, and to be able to make a decent living, has become nearly impossible. Loggers carry all the risks and the burden of responsibility for what kind of job is done, but after costs and equipment are paid for, they often have little to show for it. The pain of this injustice is personal. My father spent fifteen years of his life working with horses in the woods, and I remember his struggles because they became mine also.

A few of the works contained here were prepared by Sam Brown (a highly skilled logger-forester from middle Maine), myself and colleagues at the Hancock County Planning Commission, but it was Mitch Lansky who conducted some of the interviews, authored most of the materials, and understood better than any of us what low impact forestry was about. I am grateful to have been involved in this project and with this resultant report.

Conserving Forest Communities
by Wendell Berry

If economists ever pay attention to such matters, they may find that as the scale of an enterprise increases, its standards become more and more simple, and it answers fewer and fewer needs in the local community. For example, in the summer of 1982, according to an article in *California Forestry Notes*, three men, using five horses, removed 400,780 board feet from a 35.5-acre tract in Latour State Forest. This was a 'thinning operation'. Two of the men worked full time as teamsters, using two horses each; one man felled the trees and did some skidding with a single horse. The job required sixty-four days. It was profitable both for the state forest and for the operator. During the sixty-four days the skidders barked a total of eight trees, only one of which was damaged badly enough to require removal. Soil disturbance in the course of the operation was rated as 'slight'.

At the end of this article the author estimates that a tractor could have removed the logs two and a half times as fast as the horses. And thus he implies a question that he does not attempt to answer: Is it better for two men and four horses to work sixty-four days, or for one man and one machine to do the same work in twenty-five and a half days? Assuming that the workers would all be from the local community, it is clear that the community, a timber company, and a manufacturer of mechanical skidders would answer the question in different ways. The timber company and the manufacturer would answer on the basis of a purely economic efficiency: the need to produce the greatest volume, hence the greatest profit, in the shortest time. The community on the contrary--and just as much as a matter of self interest--might reasonably prefer the way of working that employed the most people for the longest time and did the least damage to the forest and the soil. The community might conclude that the machine, in addition to the ecological costs of its manufacture and use, not only replaced the work of one man but more than halved the working time of another. From the point of view of the community, it is *not* an improvement when the number of employed workers is reduced by the introduction of labor-saving machinery.

This question of which technology is better is one that our society has almost never thought to ask on behalf of the local community. It is clear nevertheless that the corporate standard of judgment, in this instance as in others, is radically oversimplified, and that the community standard is sufficiently complex. By using more people to do better work, the economic need is met, but so are other needs that are social and ecological, cultural and religious.

From *Another Turn of the Crank* by Wendell Berry, Counterpoint Press, Washington, D.C., 1994

Introduction

Why Low-Impact Forestry?

The Low-Impact Forestry Project (LIFP) started out in the early 1990s as a small group of landowners, loggers, scientists, and foresters who were committed to locating examples of excellent forestry, analyzing why these examples were successful, and communicating the results to the public. We were motivated to get together because so much of the forestry in our region of Maine, northern New England and the Maritimes is, frankly, shoddy. Much of the logging shows little concern about the quality of the future stand. The prime concern is getting the wood out, rather than managing a forest.

Logging machinery has gotten bigger and more powerful and is more productive at removing wood fast. But these machines can leave a big footprint in the woods. The cost of the machinery is so high that it forces loggers to cut huge quantities of wood just to pay expenses. Those loggers who take more time to be more careful are often penalized, rather than rewarded. It takes more and more forest to support fewer and fewer jobs. Such a system hurts both the forest and local communities over the long run.

Our personal observations were confirmed by state and federal studies in Maine. A survey of logging operations done by the Maine Forest Service from 1991-1993 revealed that more than forty percent of partial cuts left stands that were silviculturally understocked (with too few remaining trees to have a manageable stand). Around 57,000 acres a year of partial cuts had harvest quality (a combination of highgrading and stand damage) so poor that the Maine Forest Service found them unacceptable. Nearly half of the partial cuts had harvest quality of marginal acceptability or worse. A 1996 Forest Advisory Team that reviewed studies done on BMPs (Best Management Practices designed to prevent siltation of water) gave grades (A to F), based on compliance, to each BMP. Fifty-three percent of BMPs got a D or below.

The woodlot owners in our group did not want such shoddy practices on their land, but did not know where to find (or how to afford) foresters and loggers committed to excellent forestry. The foresters wanted to do ecologically sound management, but were constrained by landowner objectives, which sometimes were in conflict with that goal. The loggers in our group wanted to cut to high standards, but did not want to starve to do so. We identified many serious barriers (economic, political, and cultural) to wide-scale improvements in forest practices. We realized that we would have to reinvent a forestry system that acted as if the future mattered.

We were fortunate that there were experienced forester/logger/woodlot owners in our group who had demonstrated excellent forestry on their own lands over many decades. We walked through their woodlots, looked at machinery, watched logging techniques, discussed management principles, and eventually started publishing articles and setting up demonstrations of low-impact forestry (LIF).

We called these practices “low-impact” because we knew we were making a compromise. Clearly any logging leads to impacts through roads, trails, and tree removal. Low-impact forestry is a direction, not a destination. No matter where foresters, loggers, or landowners start, they can always improve their practices.

Low-impact forestry is not a substitute for wilderness. Given that our society needs some wood products, our object is to produce these with as little harm as possible. The object is to keep as many ecosystem functions of the forest intact as possible, yet still remain economically viable.

Ron Poitras, of the Hancock County Planning Commission (HCPC), came to one of these demonstrations and was impressed with what he saw. He wondered if there might be an interest in this method of forestry in his region. The HCPC did a survey of county woodlot owners that showed a strong desire to improve forest practices. Two-thirds of all respondents wanted to work cooperatively with a logger over an extended period of time to maximize economic returns and minimize environmental impacts. Nearly half of respondents would be willing to receive less for stumpage in order to improve the land's future health and productivity. More than one-third were interested in low-impact logging services.

The Planning Commission cosponsored a Low-Impact Forestry conference in May of 1997 that attracted more than 140 enthusiastic landowners, loggers, and foresters. The interest level of participants was very high, and many made it clear they wanted more follow-ups to the conference. Participants wanted more workshops on a wide range of subjects, but they also wanted to see LIF used in commercial operations.

Why this book?

To get LIF going requires committed landowners, knowledgeable foresters, trained loggers, suitable equipment, and some kind of guidelines that landowners, foresters, and loggers would agree to follow. It also requires a means of assessing logging practices to see how they compare to the desired results. Loggers need to be paid in a fair and rational way so that they are not penalized for taking time to take care. Finally, the system has a higher probability of working well with forestry associations that can help with marketing as well as other supportive functions.

From 1997, the Low-Impact Forestry Project, through the HCPC, worked to promote LIF through educational efforts--part of which included a series of publications on its web site. By 2000, enough material was on this site, covering the major elements needed in a low-impact forestry system, to put together a book.

We were on the verge of organizing landowners and loggers into an association that would have produced certified "sustainable" forest products. Just as we were doing this, a large industrial forestry company, J.D. Irving, got certified in Maine under Forest Stewardship Council standards--the same standards under which our landowners would be certified.

We saw a number of problems with continuing to seek certification for small landowners:

- Irving was using practices (clearcutting, herbicide spraying, plantations of off-site species) and logging techniques (reliance on feller-bunchers and grapple skidders) and social policies (exploitation of workers to the point that loggers and truckers were protesting) that our own members would find unacceptable.
- We had concerns about certifying based on promises, rather than on a history of sound practices. We wanted to see certification based on a proven history of passing measurable, meaningful standards.
- Ninety-six percent of certification acreage world-wide has been of big landowners (public and private)--small landowners remain at a marketing disadvantage. (For more discussion about concerns with FSC and certification, see Appendix VI.)

Due to a series of setbacks after our withdrawal from certification, including the loss of expected funding and, even more seriously, the total loss (by fire) of my house and all that was in it, the LIFP lost momentum. I had to take off time to rebuild my house and my life. Interest in LIF during this period did not fade, however. The Hancock County Planning Commission received numerous requests for information. Groups in the US and Canada have been using LIFP material (both printed and on the web) for their own organizations. The Maine Organic Farmers and Gardeners Association's Common Ground Country Fair Grounds at Unity, Maine, now has a demonstration area and annual presentations on LIF (see www.mofga.org).

Given this high interest, members of the LIFP decided that we should go ahead and publish the articles and studies from the web page into a book. Having the material all in one place makes it far more accessible than having to spend hours reading it on a computer screen. The results of our research on principles and guidelines, management strategies and technologies, methods to measure management results, economic consequences of LIF, and the options for organizing forestry associations are in this book.

This book also gives suggestions on:

- how to pay loggers,
- what should be in a contract,
- where to go for more information on machinery, and
- who to contact for more information on forestry associations.

Over the years I, and others, also published related articles, interviews, and profiles in the *Northern Forest Forum*. Some of these articles were a good fit with the material from the web page and have been included in this book. One of these articles (Chapter 1, "A Brief History...") sets low-impact forestry into an historical context and within a broader strategy for protecting forests that includes reserves and demand reduction.

The good news is that a number of forestry associations in the US and Canada are committed now to similar management principles as LIFP, and are having some success at marketing value-added products for the benefit of members. The material in this book, therefore, is not just theoretical; it is practical.

We have been heartened to discover that in a forestry atmosphere that has been heavily polarized, low-impact forestry has been a source of common ground. We have had meetings where environmentalists, foresters, and loggers have enthusiastically discussed how to make LIF work--even in the midst of bitter public debates over forestry referendums. We think we are on to something good and want to share it.

For whom is this book intended?

This book, though it covers the subject widely, is not intended as a step by step manual for the complete dummy. The point of the book is to stimulate thought and understanding. Forests and the circumstances that surround them are too variable to allow an author to anticipate every possibility. If landowners, foresters, or loggers *understand* the basic concepts, however, they can better figure out what to do. There is more than one way to meet the goals--the key is to know the goals.

This book is also not meant to be a complete academic treatise (treetise?) on the subject. It is based on years of experience of many individuals, but the subject area has much room for further exploration and discovery. For those who want to explore the literature further, there are references and, in some cases, web sites and addresses.

Low-impact forestry will not happen on the ground unless landowners request it. This book can be an important tool for helping landowners figure out what they want to have happen in their woods and how to make it happen. Landowners need to be able to communicate their plans to foresters and loggers. This book can help clarify such communications. Indeed, to the extent that landowners, foresters, and loggers have all read relevant sections, the chance for a mutually beneficial outcome increases.

You don't have to be a landowner, forester, or logger to find something of value in this book. Anyone interested in forests, forest management, or forestry policy can find some sections worthy of a read. Since most of the sections of the book were originally written as articles or studies, it is fine to browse around, and even to skip more technical sections if you wish.

Maine is the most heavily forested state in the U.S. and has one of the longest histories of continuous logging. Although most of the case studies in this book are based on the Acadian forest (Spruce-fir/northern hardwoods) of Maine, the basic concepts and principles can be applied to woodlots and timberlands in other regions if they are adjusted to the forest types and disturbance cycles of those regions.

*****

The mission of the Low-Impact Forestry Project

is to encourage:

- a long term management perspective;
- a view of the forest as an ecosystem;
- less destructive logging practices;
- high value markets for products harvested using low impact methods;
- management for multiple objectives, including social and community values; and
- productivity of the forest, broadly defined.

*****

1. A Brief History of the Ups and Downs of Maine's Forest And a Recommendation for a Three-part Forest-protection Strategy[1]

And when upon the long-hid soil the white Pines disappear,
We will cut the other forest trees, and sow whereon we clear;
Our grain shall wave o'er valleys rich, our herds bedot the hills,
When our feet no more are hurried on to tend the driving mills;
Then no more a lumbering go, so no more a lumbering go...
--from "loggers Boast," in *Forest Life and Forest Trees*, by John Springer, 1851

Part I: Devastation

Fourteen thousand years ago, Maine was covered by a blanket of ice a mile deep. By nine thousand years ago, most of Maine was covered by closed forests. The predominant species shifted as the climate changed and as the soil recovered from being scraped bare by ice. The Native American tribes certainly used the forest and forest trees for many purposes, and they may have done some burning in parts of the state, but they were not engaged in wide-scale logging. When European settlers arrived, they were impressed by what they saw.

Indeed, the first European settlers could not imagine that there might be limits to wood use. The forests seemed so incredibly vast that an economy based on wood for subsistence and export seemed assured for perpetuity. Coming from a land that had already been deforested and badly farmed out, the settlers saw the forests of New England as a kind of promised land. Native Americans had been living in Maine for thousands of years, but they had never imagined utilizing the forest on the scale or intensity that European descendants would eventually reach.

The timbers that went into this ship were hand-hewn in the woods and hauled 20 miles by team to the shipyard. From Ferguson and Longwood, *The Timber Resources of Maine*, 1960.

The wood-starved English wasted no time getting down to business. In 1605, George Weymouth brought back samples of the magnificent timbers he encountered along Maine's coast and rivers. At the time, Europeans faced a shortage of mast-trees suitable for sustaining powerful navies. The closest mast sources for England, Scandinavia and the Baltic countries, were being cut off by hostile monarchies and grasping monopolies. Yet, here

[1] First published in *Northern Forest Forum*, Vol. 4, No. 1, 1995

were white pines towering more than 150 feet, suitable, wrote Weymouth, "for ships of four hundred tons."

Maine was a prime location for ship building in both colonial and later years. In 1607, two years after Weymouth's voyage, colonists built a 30-ton ship in Sagadahoc (now Bath). By 1855, the United States had the largest merchant marine fleet in the world and Maine had supplied more than half of these wooden ships.

Commercial logging did not commence until 1631, with the first exports (mostly hand made clapboards and staves) in 1634. Due to instability of populations and problems with raids by Native Americans, a full-blown timber industry did not develop in Maine (except along the southern coast) until the early 1800s. In the 1820s, after Maine attained statehood, the industry started to boom.

By decree, during the colonial period, all sound pines over 24 inches in diameter were claimed by England with the mark of the Broad Arrow, a policy which the colonists resented and largely ignored. In 1607, all of Maine was "owned" by England. England did issue a few million acres in grants to encourage settlements. According to Great Northern historian John McLeod, "beginning immediately after the American Revolution both Massachusetts and later Maine sold and gave away their public lands to everybody and his uncle as fast as they could." At first the lands were sold for agriculture and settlement--later, in the 19th century, the land was sold to lumbermen for the timber.

In 1783, when Maine belonged to Massachusetts, less than 20% of the land was privately owned.. In 1820, when Maine became a state, 50% was privately owned. To help pay off debts, the new state started selling off its lands, mostly to timber barons. To pay for the State House, for example, Maine sold off ten townships. By 1854 only 27% of the state was public and by 1878, only 2%. The ownership of vast areas of "wildlands" by these barons (later replaced by paper companies) had a major impact on how the forest was managed. Indeed, as these big landowners carved up timber empires in the Maine woods, they dammed, and even redirected rivers, to facilitate timber drives to their mills.

Besides wood, the early settlers also found game and furs in the forests. John Smith reported bringing home 11,000 pelts from just one 1614 expedition to the Maine coast. Not only were beaver and other valuable species soon decimated, but passenger pigeons, wolves, panthers, and caribou were extirpated as habitats were destroyed and animals ruthlessly hunted.

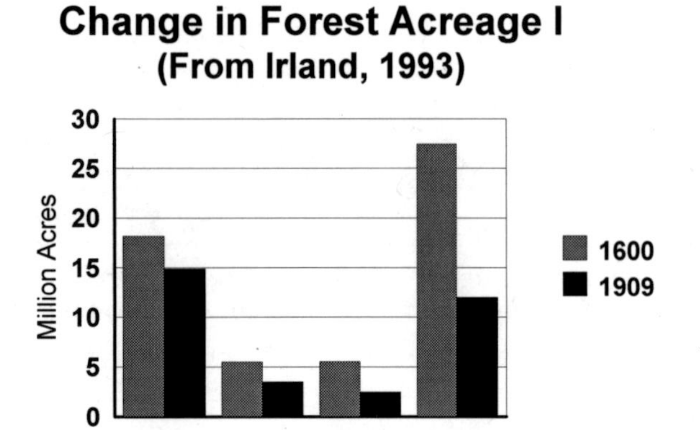

Forests were cleared and burned for agriculture. The rapidity and scale of the deforestation were staggering. By the mid-19th century, most of New York and southern New England, and millions of acres of northern New England, had been cleared, farmed, or pastured, and, in some cases, depleted and abandoned. In the north woods, many thousands of acres were cleared simply to feed the horses and oxen needed to log the forests.

Not just ships, but homes were built of wood. Maine ranked first in the country for production of saw timber up until 1800. By 1840, Maine was second only to

New York. At that time, however, the city of Bangor was the biggest producer of saw timber in the world. At first, white pine was the dominant saw timber species. By 1850, most of the easily accessible big pine had already been cut.

Observers of that time could see the trends, but they were not pushing for a sustainable woods economy. On the contrary, John Springer, in 1851, observed that, "...there is now timber enough standing in the forests, on territories through which the waters of the Penobscot pass, to maintain the present annual operations, vast as they are, for fifty successive years. After this it is thought the amount will diminish about one tenth per annum until its final consumption, when, doubtless, the pursuits of the lumbermen will give place to the labors and regards of husbandry, and the working of the various veins of mineral deposits already known and yet to be discovered."

By 1861, more spruce than pine was coming out of the Penobscot watershed. Lumber production increased up until 1909, but by then, Maine ranked nineteenth in the nation. Even bigger deposits of easily-accessible raw timber had been discovered in other states.

A major use of wood in previous centuries was for fuel. Large colonial homes might require 30 or 40 cords of firewood each winter for heat. Wood was also burned to make charcoal to fuel iron and copper furnaces. One can still see the remains of one of these smelters at Katahdin Ironworks near Gulf Hagas gorge. Before the Civil War, these furnaces might consume 150-250 bushels of charcoal per ton of iron produced. The refineries usually owned 15,000 to 40,0000 acres of forest to supply their wood needs. Wood was burned as well to produce steam for various uses, such as saw mills and steam ships. By 1850, there were 36 steam powered sawmills in Maine. Even in the late 1870s, however, most sawmills in New England depended on water power.

Hemlock bark was used to tan leather. By 1860, leather tanning was Maine's third largest industry. Historians estimate that around 9 billion feet of hemlock were cut down for the leather industry. Most of the trees, once stripped of bark, were left to rot. A tannery might use 6,000 cords of bark a year--and it took five big hemlocks to yield a cord of bark.

Wood was used to make boxes, barrels (one of Maine's first exports was barrel staves used to make hogsheads to ship rum, wine, sugar and molasses), tool handles, spools, dowels, and excelsior stuffing for packaging. In 1860, Charles Foster opened a toothpick factory in Strong, Maine. That year he turned out 16 million white birch toothpicks so refined people could clean their teeth after meals.

And then, industrialists started making paper from trees. The first production, where ground wood was mixed with rags, was in 1868, in Topsham, Maine. By 1880, several companies had begun producing paper solely from wood pulp. By 1889, a number of mills started using sulfite processing for pulp production. A new industrial era was dawning. Deforestation proceeded at a frightening rate.

"If it were wise to live today, as if there were no tomorrow, this [destructive cutting] might be excusable; but with the existing impression that the country will remain after we are gone, and that our children will need, and will be thankful for the heritage unimpaired, it seems like the folly of madness, not the wisdom of wise men, to pursue the course we are now pursuing."
Editorial, Portland Eastern Argus, 1883

By the end of the 19th century, nearly all the virgin forests in the region were gone. Industrialists dependent on water power began complaining about severe fluctuations of water levels and of siltation fouling their mills. They started demanding protection of headwaters. In 1901, Francis Wiggin addressed the State Board of Trade in Rockland calling for the establishment of forest reservations by the right of eminent domain. He suggested starting with the regions around the Rangely Lakes area, Moosehead Lake, and the West Branch of the Penobscot River.

While New York got the Adirondacks in 1894, and New Hampshire the White Mountain National Forest (after the Weeks Act in 1911), the Maine legislature took no action to create reserves or parks in the north woods. Percival Baxter, many years later, had to personally buy land around Katahdin and then donate it to the state to create a state park.

Even some members of the forest industry expressed their concerns about forestry trends. In 1892, William Russell, the president of the American Paper Makers Association, warned his group that "Certainly we shall keep on denuding the forest for as we are turning almost wholly to wood as a fiber...we are drawing on the forest rapidly. I hope that some wiser way of cutting our timber in this country will be devised so that we shall not...see the end of our spruce forests. We must either of our own volition, or by some government control, prevent the destruction..."

In 1907, a bill was sent before the legislature to set diameter limits on the cutting of pine and spruce. This controversial bill was sent to the Maine Supreme Court to determine if it was a "takings." The Court, in an historic ruling on March 10, 1908 declared that regulating forestry practices was indeed legal. Assuming that Maine would act upon this judgment, President Theodore Roosevelt singled out Maine for its "wisdom and foresight" at a governors' conference on conservation that year. The 1908 decision did serve as a precedent for other states, but Maine virtually ignored the ruling. Attempts to pass a forestry bill in 1909, 1913, 1915, 1917, and 1919 were all defeated in committee. The limits to cutting, by default, were left to landowner objectives and the market.

"We think it a settled principle, growing out of the nature of a well-ordered society, that every holder of property, however absolute and unqualified may be his title, holds it under the implied liability that his use of it shall be so regulated that it shall not be injurious to equal enjoyment of others having an equal right to the enjoyment of their property, nor injurious to the rights of the Community.

"While it might restrict the owner of wild and uncultivated lands in the use of them, might delay his anticipated profits, and even thereby cause him some loss of profit, it would nevertheless leave his lands, their product and increase, untouched and without diminution of title, estate or quantity. He would still have large measure of control and large opportunity to realize values. He might suffer delay but not deprivation."

Maine Supreme Court, March 1908

Nature strikes back

With the heavy cutting and other human-caused disturbances came greater forest instability. Soon after settlement, huge fires consumed thousands of acres of forests. The fires tended to follow heavy cutting (indeed, moist, dark virgin forests often acted as barriers to put fires out). Fires swept through the coastal forests, and then, as logging moved north, so did the fires.

In 1825, the "Miramichi" fire (so named because it coincided in time with an unconnected fire in New Brunswick) burned an unbelievable 832,000 acres. The fire in New Brunswick burned an estimated *two million* acres. The Katahdin area had another major fire of 20,000 acres in 1884. Nineteen years later, in 1903, 84,500 acres burned in the Katahdin region and around

270,000 acres of forests burned in the state. That same year, more than 637,000 acres burned in the Adirondacks.

Burned land on Chase Stream Township from Colby, *Forest Protection and Conservation in Maine*, 1919.

These massive burns were but a prelude to even greater destruction. At the turn of the century, the larch sawfly decimated the tamaracks of the state. From 1911 to 1919 a severe spruce budworm outbreak swept through millions of acres of forest, killing an estimated 40%, or 27.5 million cords of the spruce and fir in Maine, and 225 million cords in northeastern North America. In some areas, the dense fir thickets, which grew to fill the gaps left by heavy logging of spruce, suffered complete mortality. Even in older stands, not just fir but spruce and hemlock were impacted.

Recovery

The clearing, the logging, the fires, and insects and diseases caused a major reduction in both acreage and volume in the northern forests. Conservationists at the turn of the century were decrying much of the destruction as needless waste. They saw lumbermen cutting high on the stump and cutting off tops with much good wood going unused. The slash was creating tremendous fire hazards. Waste from sawmilling, they insisted, could be used to create boards (with glue) or for pulp. Their arguments were given force by support of the renowned forester, Austin Cary, who advocated setting large diameter-limits (leaving all trees with diameters smaller than the set limit) and doing more selective cutting.

In 1902, Maine Forest Commissioner, Edgar Ring, stated that, "I am satisfied that the supply of available material to carry on our present pulp and lumber manufacturing establishments, and such others as may be built as time goes on, is sufficient and ample..." By 1919, the Maine Forest Service changed its tune. Commissioner Forrest Colby stated in his annual report, "It is a fact not to be disputed that we are cutting off our forests today much faster than they are being reproduced, and we have been doing this for years...That we have wasted our forest is a matter of common knowledge."

Colby recommended a three-part strategy for protecting the forest:

1) Improved fire control,
2) More state ownership, and
3) Reducing logging waste and planting over "wastelands."

The first of these strategies to be taken seriously in the state was fire control. The 1903 fires sent a jolt through the legislature, and in 1905, Maine established the nation's first fire towers. In 1919 a law was passed that gave a 20-year tax break for those who planted trees on cutover land. In 1921 a law was passed encouraging the purchase of state lands and the creation of "auxiliary" state lands. The auxiliary lands were privately owned, but subject to strict diameter limits designed to grow big pine and spruce. This law was amended in 1923 and 1929, and then finally repealed in 1933. Very few people took advantage of the tax breaks. There was more incentive to cut the wood.

Forest industry spokespeople cannot resist comparing the forests of today with those at the beginning of the century. They like to imply that the increased volume and acreage of wood is due to their benign management. While some companies did, under the influence of Austin Cary and other reformers, start using diameter-limits for cutting their spruce,[1] and did cut lower on the stump and higher up to the tops of trees, such changes were not the primary cause for forest recovery.

What brought back a new forest was not enlightened forestry, but the abandonment of farms, a decline in logging, and improvements in forest fire control. Historian David Smith estimates that the cut after 1930 was less than 1/3 of that in 1909.

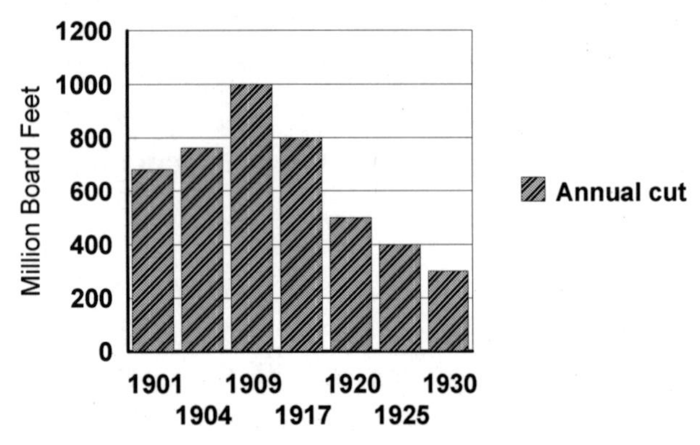

From Smith, 1972

According to Smith, the cut did not rise above this level again until after WWII.

Part II: Highgrading

If there has been a constant theme in resource use, it has been highgrading. Farmers and foresters have been attracted to the best soils, the biggest trees, and the easiest access. As the poorer, rockier soils in New England got depleted, farmers moved West to much more productive lands. They were able to take advantage of government-subsidized rail and barge systems to transport their farm goods to population centers. As New England farms were abandoned, starting in the 19th century, they grew back into forests.

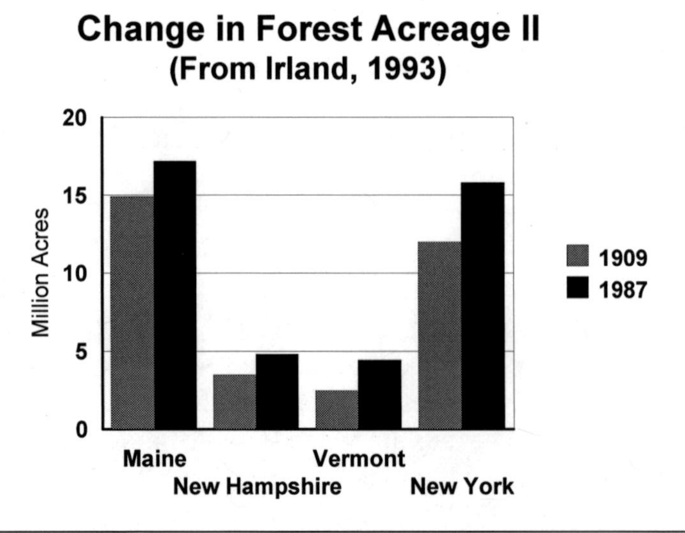

Loggers looked first for the biggest trees near lakes, rivers, and large streams. To the benefit of timber companies, Maine has an abundance of water bodies suited to transporting logs to mills (with the aid of dams). Maine also has long winters with reliable cold and snow coverage. This allowed transport by

[1] Diameter-limit cutting is not necessarily good silviculture. It does not deal with stocking or quality. It is basically a cutting rule that can, in many circumstances, lead to highgrading, especially when the limits are low.

animal-drawn sleds. Because of these natural advantages, Maine companies were able to avoid making large investments in transportation infrastructure--until the latter half of the 20th century when river drives were phased out and eventually banned.

Pulpwood sluice on Pogy Mountain. From Colby, *Forest Protection and Conservation in Maine*, 1919.

As transportation and logging technology improved, loggers were able to go farther afield in search of big timber. Highgrading was thus both by the landscape (the best, most accessible forests) and by the site (the biggest trees). Later cutting was at less productive sites and with less impressive trees.

By the middle of the 19th century, loggers were already scrambling to find the pine in more difficult locations. John Springer, in 1851, wrote of the techniques the lumberjacks used to get big timber off of steep slopes and mountain sides. Although sluices were preferred, loggers sometimes lowered the big trees down with block and tackle, or even shoved them off of cliffs.

After most of the big trees had been skimmed from the Northern Forest, loggers, taking advantage of new transportation networks, moved to more productive regions. The Erie Canal, opened in 1823, freed up wood markets from the Lake States to flood into eastern markets. The Panama Canal, in 1914, did the same for the Pacific Northwest. The Great Northern Railroad, in 1885, also allowed Western wood to enter the East at competitive prices. After 1909, the lumber industry in Maine went into decline. There was better wood available at better prices from elsewhere. By the late 19th century, for example, big mast trees, even on ships built at Bath, were coming from Oregon.

As wood became more scarce (and more expensive), industrialists came up with substitutes. By the 1860s, iron began replacing wood in ship building. Changing hide sources, new technologies, and industry consolidation caused a rapid decline in the use of hemlock for leather tanning in this region in the early 20th century. Cardboard replaced wood for construction of boxes. Coal took the place of charcoal for iron furnaces and steam mills. City building were increasingly built with steel beams, concrete, and bricks. Oil also allowed the substitution of machines for animals in the woods, freeing up thousands of acres of pastureland to return to forest.

Even as abandoned farmland was growing back and as the lumber and tanning industries declined regionally (having moved on), another force appeared to save the forest--the Great Depression. The big American boom went bust. Demand declined. Growth in the forest was greater than cut. The battered, bugged, and burnt forest started to recover.

After World War II (with the exception of 1947), new technologies--from tank trucks to airplanes--made sure that fires burned far fewer acres. Hunting and trapping regulations allowed for the return of some of the endangered game animals--including the moose and beaver--to the new forest. The northwoods showed its resilience.

Fire Control in Maine (1910-1969)
(Ferguson and Kingsley, 1972)

Decade	Number of fires	Total acres burned	Average acres per fire
1910-1919	1,121	229,784	205
1920-1929	1,744	275,251	158
1930-1939	2,516	282,360	112
1940-1949	4,552	337,942	74
1950-1959	5,903	107,631	18
1960-1969	5,292	42,227	8

New devastation

What came back, however, was not quite the same as what had come down. The size, structure, and species ratios had changed. But this new forest was just fine for the paper industry, which, after the depression and war, began to rebound with the forest. New technologies of chainsaws, skidders, pulptrucks, and timber harvesters allowed the paper industry, which bought out millions of acres of forests from the lumber barons, to reach new deposits of timber or more completely utilize old ones.

Technological change did not improve the quality of forest species or trees; it allowed the companies to better use what was available. First loggers cut pine and spruce to a 12-16 inch diameter limit because that was the sawmill market limit. Forests generally recovered when these higher limits were used. With the pulpwood markets opening up for spruce and fir, however, cutting could go to much smaller diameters and be much more intense.

Pulpwood operation From Colby, *Forest Protection and Conservation in Maine*, 1919.

By 1972, the USDA Forest Service lamented that hardwoods had been so repeatedly highgraded that growth rates and quality were abysmal. "So little hardwood timber in Maine is in managed stands that for all practical purposes one can say that generally in Maine hardwoods are not managed." By 1986, the Maine Forest Service declared that 43% of all hardwoods were "unsound." A new industry--biomass for steam and electricity--evolved (with government and environmental group help) that could use this low-grade material--branches, tops, and all. Whole-tree clearcutters now had a market for all fiber.

In the 1970s, another spruce budworm outbreak erupted in full after sputtering along since the 1950s. Once again, abnormal concentrations of balsam fir had filled in the gaps left from the last harvest/budworm spree. This time, however, the industry, with government help, was able to spray millions of acres with broad spectrum chemical insecticides. And this time, with more roads and more machines, big landowners "salvaged" with a vengeance. New technologies capable of better utilizing smaller-diameter trees encouraged the growth of a new lumber and stud-mill industry, creating more markets to absorb the massive cutting.

Rolling clearcut from the 1980s. Photo by Chris Ayers, first published in *Beyond the Beauty Strip* by Mitch Lansky, 1992

The clearcutting was far more widespread and severe than even the fires earlier in the century. Adding insult to the injury, companies using heavy machinery rutted and compacted the soil, removed whole trees, and topped off the disaster with a blanket of herbicides to keep down pioneer species. The industrial landowners looked at what they had done (it was visible by satellite from outer space) and declared that it was good. They called it "intensive management."

A USDA Forest Service inventory in 1995 showed a decline in total volume of the forest inventory of Maine. The decline was due primarily to a combination of heavy cutting and a budworm outbreak in four counties--Aroostook, Piscataquis, Somerset, and Franklin. Red spruce was cut at a higher rate (as a percentage of inventory) than any other species, despite recommendations by experts to spare the spruce and concentrate the cut on the far more vulnerable fir. With the exception of Aroostook, the decline in these counties was for hardwoods as well as softwoods. According to figures from the USDA Forest Service, between 1982 and 1994, there was an 80% or more removal of volume on 2.2 million acres (over 3,400 square miles) in Maine. During that same period, the acreage of saplings and seedlings increased by 1.2 million acres.

Once again, the combination of demand from people and bugs had exceeded the growth capacity of the forest. In addition, the forest is now subject to air pollution, climate change, introduced insects and disease, development, and other stresses that may make recovery more difficult this time around.

As happened at the turn of the previous century, citizens are demanding reserves and regulations. We actually have regulations in Maine now, protecting, to some extent, special management areas such as riparian zones, deer yards, eagle nests, or high-altitude zones. Regulations outside these zones (which make up a small minority of the forest) have more impact on the way clearcuts are arranged across the landscape than on sustainable levels of cut. All we have to save us from overcutting, at this point, are, once again, market forces and landowner objectives.

Recovery?

New markets for hardwood pulp, biomass, lumber, and exports have replaced the markets that declined earlier in the century. Industrial landowners have reduced the cut on their own lands (and many have sold their lands off to investor groups), increased purchase of wood from non-industrial landowners, and increased importing wood from out of state. Maine is now a net importer of wood (mostly pulpwood) although, ironically, it is a net exporter of quality sawlogs. Forty percent of the spruce-fir sawlogs in the state are exported to Quebec sawmills. The "waste" (edgings, slabs, and sawdust), is imported back for pulp. The mills are also substituting more hardwoods into their pulp mix. Indeed, more hardwoods than softwoods are now being cut for pulp.

Stream drive--Working out a side jam. From Colby, *Forest Protection and Conservation in Maine*, 1919

With markets more globalized, people can live with the illusion that there are no limits. Indeed, there are a few "fiber mines" left that have not yet been fully exploited. Due to poor infrastructure and political instability, for example, there is still a huge forest "resource" available in Siberia. If that gets opened up to the global economy, it might do for over-exploited timber regions what Minnesota or Oregon did for Maine many decades ago. People will be able, temporarily, to continue high levels of consumption without seeing the consequences at a local level. But if these trends continue, Siberia will suffer the same fate as other exploited regions.

The timber industry (with some government support) has been offering another fix for the demand problem--intensive management. They point to the graphs that show rising demand and conclude that the answer is rising supply. The best way to increase supply, they argue, is through short-rotation plantations, herbicides, and pre-commercial thinning of clearcuts. They wouldn't mind if we, the public, subsidized these activities--which are, after all, for our own good.

The intensive management fix assumes that demand (as projected by industry or government officials) is sacred and that we have a moral obligation to meet it. It assumes that this form of management will yield as promised and that the public, which hates clearcutting and herbicides now, will grow to love them. It also assumes that the world will either end in the year 2050 or that what comes after that year is not worth pondering.

Unfortunately, the intensive management fix ignores the basic question: How can we have ever-increasing fiber demand in a world with limited forests? Even if foresters found a way to double forest production over the next 50 years, if population and per-capita consumption increase as well, all this growth will be devoured--and then what? There are, after all, limits to yields from intensive management, just as there are limits to forest acreage. Do we want to meet the limits of the forest after vast acreages have been converted to simplified landscapes growing low-value commodities or do we want to meet the limits while we still have options and something resembling natural forests? We could wait for another depression, or we could, as a society, consider ways to deliberately reduce demand from our forests.

Part III: Reducing Demand

Our experience earlier in this century demonstrated that forests (of a sort) could grow back if the cut is reduced. The way the cut was reduced, however, did not truly solve the problem. Population increased by multiples. The problem of non-sustainable farming was foisted onto the Midwest. Non-sustainable forestry went from the Lake States, to the Pacific Northwest, and is now most concentrated in the Southeast. Non-sustainable energy use has been shifted to fossil fuels. Indeed, some of the substitutions for wood, such as steel, plastic, or concrete, can use up more energy and cause more environmental damage than cutting trees (if the trees are cut in a low-impact manner).

In 1994, a group of activists, organized by the Rainforest Action Network and the Turner Foundation, met in Tomales Bay, California with the purpose of saving the world's forests by confronting overconsumption of wood products. Their overriding principle was that this should be done without creating serious new environmental problems. They determined that North Americans could reduce wood-product consumption by 75% if our society systematically used the right tools. This sounds rather drastic until one learns that Americans consume more than six times the wood and paper products per capita as does the rest of the world on average.

The group looked at the major industrial wood use sectors:

- Paper and packaging;
- Construction; and
- Pallets/shipping uses/finished wood products.

They then looked at how each of these sectors could reduce raw wood use by using the following strategies:

- Decreased per capita consumption;
- Increased re-use;
- Increased recycling;
- Increased efficiency in production;
- Increased use of alternative materials; and
- Product durability and design improvements.

They then asked what tools for change could bring about the desired goals and looked at:

- Regulations and legislation proposals/enforcement;
- Public education and advocacy;
- Fiscal and economic incentives; and
- Technical solutions.

The group published a 70-page pamphlet[1] that details how this campaign can work, with examples of organizations that are already using some of these tools. For example, wood could be reduced in housing by more efficient use of framing materials, use of salvaged materials,

---

[1] For a copy, write to Wood Use Reduction Campaign, Rainforest Action Network, 450 Sansome, Suite 700, San Francisco, CA 94111. Send $10.

substitution with more benign materials (such as rammed earth in some parts of the country), and even changes in use. We now have the dubious trend (due to the breakup of families and communities) of more buildings and material goods for smaller households. Thus, part of the solution to overconsumption is social, rather than simply technical.

Forest cutting for paper use can be reduced by such measures as more efficient use (and reuse) of office paper, ending junk mail and superfluous packaging, sharing of newspapers and magazines, recycling, and substitution of other materials (such as agricultural waste, hemp, or kenaf).

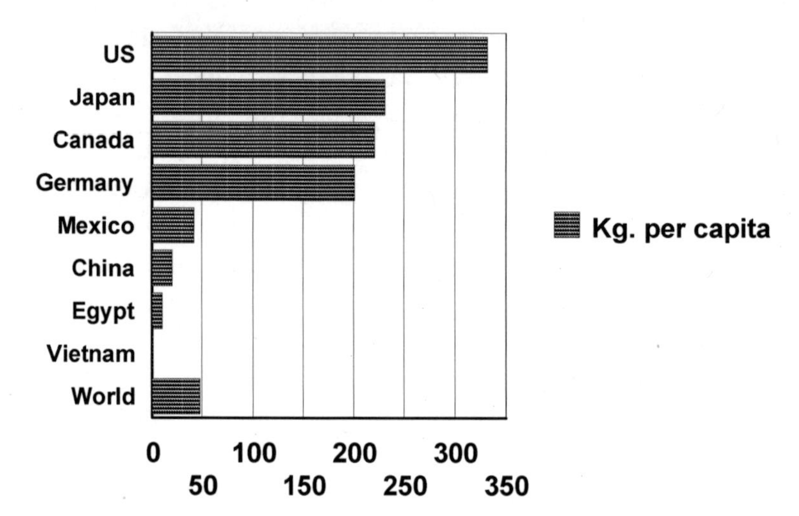

Government policy is crucial for achieving raw material use reductions. The government, for example, can set the standard for lowering waste and increasing efficiency in its own operations. It can also set policies to stop subsidizing waste in the private sector. One of the most powerful incentives to stop waste is the market. Too much of our government policy is used to subsidize waste and artificially cheapen forest products. As long as we have throwaway wood and paper products, we will have throwaway forests. If there is full cost accounting that gives a more accurate value of the product, prices would go up. When prices go up, waste goes down. We need to examine such subsidies as tax breaks, direct payments, lax regulations, and protected market domination (that allows for squeezing suppliers and workers) to see if the impacts are perverse to both society and the forest. Because markets are global, rather than local, these reforms need to be global, otherwise investment capital will flow to countries and regions where companies have the lowest costs, because they get the best government deals.

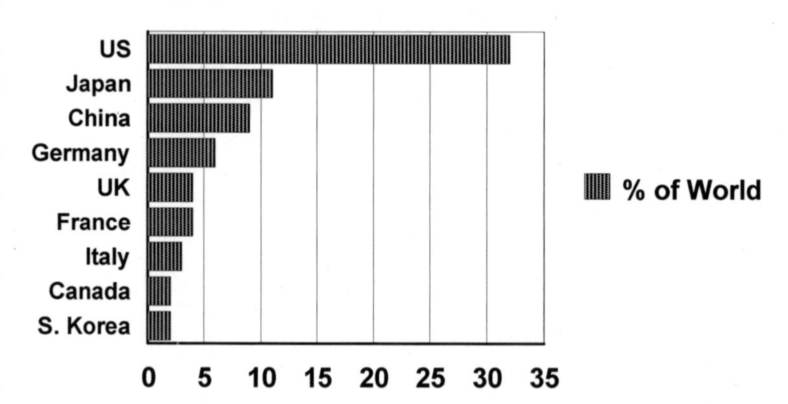

Source: Carrere and Lohman
Pulping the Sourth: Industrial Tree Plantations and the World Paper Economy. Zed Books, 1996

Conclusion

If we truly want the northern forest to recover to sustainably provide both ecological and social benefits, we will need a new, more complete strategy. University of Maine professors Robert Seymour and Mac Hunter have proposed a three-part strategy that includes reserves, "New Forestry," and "intensive management." Leaving the name "Triad" to Hunter/Seymour, the new trio (treeo?) of strategies would be reserves, low-impact management, and demand reduction.

Unlike the Triad, where biodiversity is sacrificed in one part for the sake of temporarily maintaining it in the other parts (because intensive management can not perpetually increase yield to meet perpetually increasing demand), all three elements of the Treeo strategy work toward forest recovery. Even the "Treeo" is inadequate, however, if our society does not deal with acid precipitation and global warming.

The "Treeo" Forest Protection Strategy

- ✓ Wild, unmanaged areas are a baseline against which to measure our management impacts. They are also places that allow natural processes and where we can study forest dynamics. A complete *reserve system* ensures that all habitats and species are protected--including ones that are sensitive to human encroachment or that require old growth.
- ✓ Given that our society needs forest products, *low-impact forestry* supplies them in a way that minimizes known impacts to the forests from which the products come. Managed areas can thus complement, rather than isolate, reserves.
- ✓ *Demand reduction* ensures that any local reduction in cut, due to reserves or cutting restrictions, does not get translated into greater environmental or social damage elsewhere. Demand reduction deals with waste and inefficiency, but also addresses the trend of unlimited growth of consumption.

We have seen, from this brief historical sketch, that reduced cutting can allow forests to grow back. The rest of this book will set out to show how low-impact forestry can supply forest products while still supplying biological and social benefits. The case for reserves has been made in many other publications and was made by conservationists a century ago.

Unless demand reduction (including an end to geometric population growth) is central to our strategy for sustainability, we are just fooling ourselves and robbing our descendants. Forests can sustain themselves quite well, and have done so for millions of years. Sustainable forestry is an extension of human society. If the human society is not sustainable, the forestry will not be sustainable either.

This story of forestry exploitation by European immigrants begins 400 years ago with Weymouth's voyage to the coast of Maine. There are red spruce trees in Baxter State Park that are more than 400 years old. This history, therefore, is shorter than the potential life-span of the dominant tree of the region. It has been said that a forestry system can not be adequately tested as sound unless it has stayed productive for more than three rotations. For old spruce, this implies a period of more than a thousand years. If we continue our current trends of consumption and pollution, what are the odds that our society could last a thousand years more?

Sustainability alone is not a sufficient goal for forestry policy. Questions about control of the forest and of markets must be raised as well. A "sustainable forest" that benefits only a small handful of absentee landowners is hardly a desirable outcome.

Even this issue was anticipated by writers of previous centuries. John Springer, in 1851, saw the possibility of a great future for the city of Bangor, which was, he believed, "surrounded by resources of wealth altogether beyond any other town or city in the state..." But he warned that:

"Of one great disadvantage, which must retard her progress, mention may be made, viz., capitalists abroad own too much of the territory on her river. A judicious policy in business must be steadily pursued, else she may only prove the mere outlet through which the wealth of her territory shall pass to other hands..."

Sources

Colby, Forrest, 1919. *Forest Protection and Conservation in Maine.* Maine Forestry Commission, Augusta.

Ferguson, Roland and Neal Kinglsley. 1972. *The Timber Resources of Maine.* USDA Forest Service, Upper Darby, PA.

Ferguson, Roland and Franklin Longwood. 1960. *The Timber Resources of Maine.* USDA Forest Service. Upper Darby, PA.

Griffith, D.M., and C.L. Alerich. 1996. *Forest Statistics for Maine, 1995.* FIA Unit. NEES Res. Bul. NE-135, Radnor, PA.

Holbrook, Stewart. 1961. *Yankee Loggers: A Recollection of Woodsmen, Cooks, and River Drivers.* The International Paper Company. New York.

Irland, Lloyd,C. 1994. *Wildlands and Woodlots: The story of New England's Forests.* Working Draft, cited with permission from author.

Irland, Lloyd. C. An address on the resiliency, recovery, and challenges of the Maine woods given at the Forest Products Society 1994 convention in Portland.

Jacobson, George and Ronald Davis, " Temporary and transitional: the real forest primeval, the evolution of Maine's forests over 14,000 years, " reprinted in *The Northern Forest Forum,* Vol. 1, no. 3, 1993.

Judd, Richard W., 1997. *Common Lands, Common People: the origins of conservation in northern New England.* Harvard University Press., Cambridge, MA.

Lansky, Mitch. 1992. *Beyond the Beauty Strip: Saving What's Left of Our Forests.* Tilbury House Publishers. Gardiner, ME.

Lansky, Mitch. 1998. *The 1995 US Forest Service Inventory of the Maine Woods: What does it show?*

McLeod, John E., "A brief account of how Maine's public lands were sold or given away." Reprinted in *The Northern Forest Forum,* Vo. 1, no. 6, 1993.

Ring, Edgar, 1904. *Report of the Forest Commissioner.* Maine Forestry Commission, Augusta.

Seal, Cheryl, 1992. *Thoreau's Maine Woods Yesterday and Today.* Yankee Books, Emmaus, PA.

Smith, David C., 1972, *A History of Lumbering in Maine 1861-1960.* University of Maine, Orono.

Springer, John, 1971 (originally published in 1851), *Forest Life and Forest Trees,* New Hampshire Publishing Company, Somersworth.

Chapter 1 Endnote

How has Maine's forest changed?

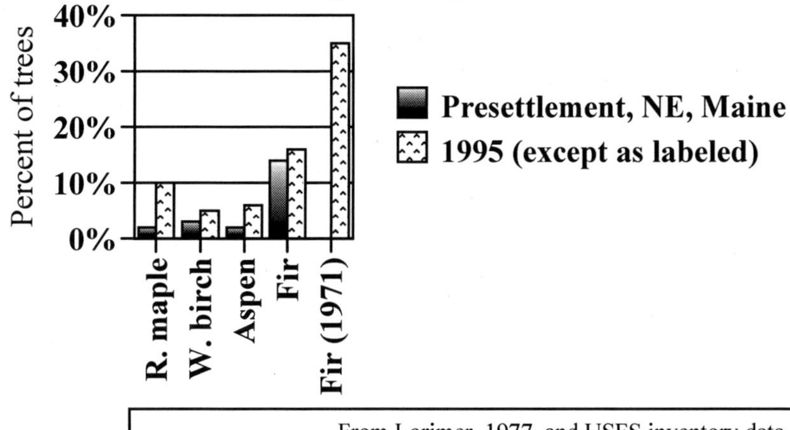

Although Maine's forest area has bounced back to a considerable degree, the new forest is different from the presettlement forest. For one, forest types have changed. Since 1959, the spruce-fir forest type in Maine went from 8.4 million acres to 6 million acres--it is no longer the predominant forest type in the state. Disturbance-adapted hardwoods, such as red maple, white birch, and aspen, are far more prevalent than they were in the presettlement forest of northeastern Maine (as reconstructed from early witness-tree data by ecologist Craig Lorimer in a 1977 study).

Balsam fir, another short-lived species that responds well to disturbance (although, unlike the hardwoods, it is shade tolerant), made up 16% of the forest in the four-county region of northeastern Maine in 1995, but this was after a major crash in inventory, due to a spruce budworm outbreak. In 1971, fir made up 35% of all growing-stock trees in the region. Fir is rebounding--it makes up a third of all 1-to-2-inch trees in the state, much to the appreciation of the spruce budworm, which happens to like basam fir.

The biggest change since presettlement times is in age-class structure. Lorimer's reconstruction showed that 59% of forest stands had grown more than 150 years from the last catastrophic disturbance (such as fire, wind, or heavy cutting). By 1995 the percent of such stands was insignificant. Whereas older forests tend to have bigger trees, more dead wood, and more vertical stratification and complexity, younger stands tend to be far more simplified in structure. This simplified structure means fewer habitats for specialist species,.

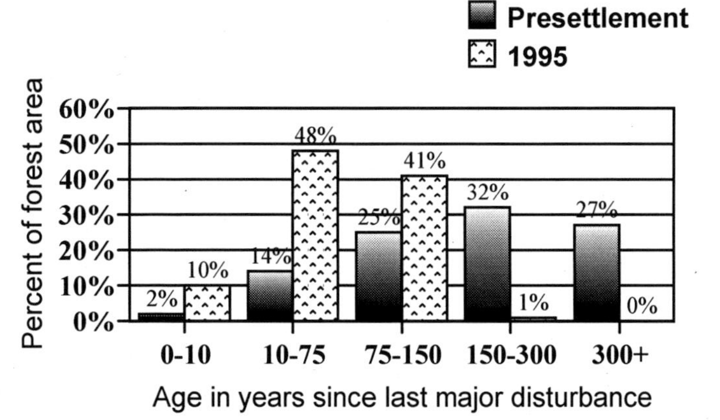

2. Principles, Goals, Guidelines and Standards for Low-Impact Forestry

I. Introduction

Too often, landowners who care deeply about their woodlands have been dismayed at the aftermath of cutting on their land. The logger promised to cut the lot "selectively"--and he did. He "selected" the best trees and left the rest, with little care for the future. The resulting stand is so sparse that it offers little shade. Runoff from rutted and compacted trails (which seem to run everywhere) and from poorly-planned crossings have silted up the once-beautiful trout brook. Many of the remaining trees (most of which are small diameter or culls) are bruised, broken, or bent. The yard is a large, muddy mess, with unsightly slash piles. Understandably, after such an experience, many landowners are discouraged from having any more logging done to their land.

There is an alternative--low-impact forestry (LIF). Low impact forestry reduces known harmful impacts so that after the cutting is done, there is still a functional forest. The landowner, with the help of a forester, plans for the long term, not just one cut. The residual stand not only functions like a forest, it *looks* like a forest. There are enough trees, including some with large diameters, to ensure that the forest floor is shaded. Very few trees are damaged. Indeed, after the cut the average tree quality is higher--the logger removed high-risk, low-quality trees. Trails are relatively narrow and unobtrusive. They are favored by those who want to cross-country ski or hunt. The yard is small with minor soil disturbance. The stream remains clear and cold, and the trout seem happy. And so is the landowner.

For LIF to work, good communication between the landowner, forester, and logger is essential. If the logger and landowner do not have common expectations and do not communicate well, the job will not come out the way the landowner wants. Often loggers are responding to immediate economic pressures that reward more volume for lower costs, while many landowners have woodlots as a sideline rather than as a primary income source. This chapter is intended to give landowners an understanding of the reasons behind LIF so that they can better communicate their objectives. Indeed, for the system to work well, it is best if landowners have their management objectives in writing. This document lists recommended guidelines and standards for foresters and loggers that, if followed, would help meet the LIF objectives.

II. Principles and Goals

With low-impact forestry, the whole forest is considered, not just its value for pulpwood or sawlogs. Foresters must look at the crowns (tree foliage), the trunks, the roots, the soil, the water, forest stand structure, and the distribution of wildlife habitat across the landscape. For a forest to be "functional," it must have all the required parts, and all the required processes. Forests, however, are always changing due to human management and natural *disturbances* (such as wind, insects, diseases, or fire). LIF foresters must be prepared to accommodate this change so that over time, the parts and processes are still functioning in the forest landscape.

Crown

Principles. Looking up in a forest, one should see the crowns of individual trees forming the forest *canopy*. The crown includes the limbs, twigs, and leaves. The leaves are where photosynthesis takes place, capturing the carbon from carbon dioxide to build the fibers in wood. They are also where trees release water, drawn from the soil by the roots, back into the atmosphere. This helps to regulate both the water table and the climate. Tree canopies (both living and dead) are habitats to numerous species of birds, insects, spiders, and even lichens. Damage to tree crowns, especially to the leader, can impact both the form and health of the tree. Excessive crown damage can lead to tree death, by drastically reducing the energy production from leaves or needles required to sustain the tree. As people discovered after Maine's ice storm of 1998, however, to kill a tree, such crown damage has to be severe.

The degree to which the canopy is *closed* (where the foliage of one tree tends to merge into the foliage of the next), can have profound effects on forest productivity, tree quality, forest regeneration, wildlife habitat, *windfirmness*, (the ability of trees to withstand falling over in heavy winds), and water quality. The more the canopy is filled with foliage, the more efficient the use of the growing space and the higher the productivity of the mature trees. As the canopy opens, more energy goes to the *understory*--the developing growth below the canopy.

Degree of canopy closure also affects the quality of many tree species. Increased sunlight from heavy cutting can lead to epicormic sprouting (sprouts coming off the trunk), leading to lowered quality. When the canopy is open, more growth goes onto branches. Trees therefore can have more knots, shorter boles, more taper, bigger crowns, and thus worse form for lumber. With higher degrees of crown closure, tree growth goes up rather than out. Increased density leads to trees with longer boles and smaller crowns. Lower branches tend to self-prune, yielding more limb-free logs.

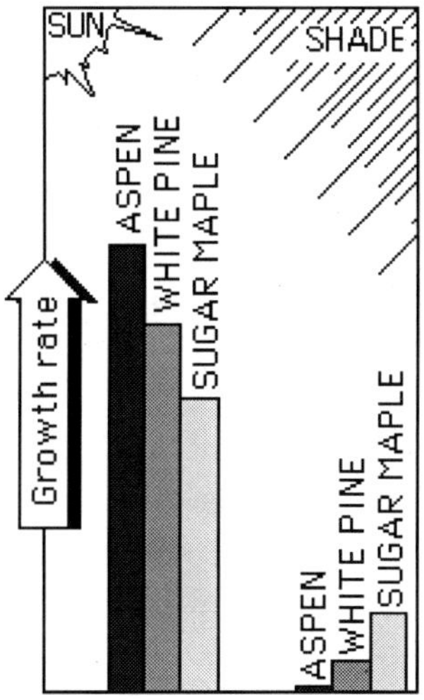

Degree of shade influences the regenerating stand. Some trees, such as birch and aspen, are adapted to heavy disturbances, such as fires, and thrive under direct sunlight. These trees, along with plants such as raspberries and pin cherries, are *shade intolerant*--they must have direct sunlight. The majority of our most valued tree species, however, are *shade tolerant*--they are adapted to growing under some degree of shade. Species such as red spruce, hemlock, or sugar maple can grow under a dense shade. White pine, yellow birch, or ash do well under partial shade and are considered to have *intermediate* tolerance.

Openings in the canopy are called "*gaps*." Gaps, depending on their size, produce different types of habitat. If they are small, they might just stimulate the growth of existing shade-tolerant seedlings and saplings, leading to a *stratified canopy* (a canopy with multiple layers) and *uneven-aged* stand structure (having three or more age classes). Stratified canopies have a high diversity of invertebrates.

If the gaps are bigger than two tree heights in diameter, the forest floor may be bathed in sunlight, encouraging shade-intolerant species, starting the process of *succession*. Since shade-intolerant species cannot grow under their own shade, more shade-tolerant species tend to succeed them. Young trees and shade-intolerant trees are found in *early-successional* stands, which are *even-aged* (having only one or two age classes of trees). Mature and old intermediate and shade-tolerant species are found in *late-successional* stands.

In Maine's *presettlement* forest (the forest before Maine was settled by Europeans), most of the northeastern portion of the state was in relatively-closed-canopy stands of more shade tolerant species. Large, severe disturbances (such as fire or windthrows) happened on a given acre hundreds of years apart--early-successional stands only represented a small percentage of the landscape. Stands dominated by seedlings and saplings may have represented only 2% of the landscape. Smaller disturbances led to an uneven-aged structure and stratified canopy, with most stands having trees older than 150 years.[1]

In some counties, such as Aroostook, Piscatiquis, Somerset, and Washington, seedlings and saplings now make up around 30% of the landscape with most trees under 80 years. Mature saw-timber stands (stands dominated by trees large enough for saw timber) are a minority in these counties.[2]

Goals. Low-impact practitioners strive to manage for *well-stocked* (having optimal spacing for productivity and quality) stands with minimal crown damage. LIF foresters favor, over time, late-successional species and canopy structures. Large gaps and early-successional stands should be a minor part of the landscape.

Trunks

Principles. Tree trunks, or boles, hold up the canopy to catch the sunlight. The trunk carries food made from the leaves down to the roots through the inner bark *(phloem),* and the water and minerals are taken by the roots up to the leaves through the sapwood *(xylem).* A thin layer (one cell thick) called the *cambium* is where the tree trunks grow. Tree trunks are most vulnerable to damage when the sap is flowing, in the spring and early summer. Damage to tree trunks leads to a response called *"compartmentalization"* to isolate the wound. This takes up precious energy, slowing growth. If the damage is severe, this can cause extensive rotting and decay, lowering value for timber products and eventually killing the tree.

Tree damaged by mechanical harvester. Photo by Mitch Lansky

As the diameter of the trunk increases, its value for timber and wildlife increases as well. The highest-quality trunks are formed in relatively-closed canopy stands where limbs self-prune. There can be a major jump in economic value as a tree becomes big enough for sawtimber, rather than pulpwood. A softwood sawlog starts at 8-10 inches and a hardwood sawlog starts at 10-12 inches, while prime veneer starts at 16-18 inches. The difference in value between a 12-inch sawlog and an 18-inch veneer log can be 400-500%.[3] One therefore grows more value per acre per year on trees than on saplings, on sawlogs than on pulp, and on veneer than on sawlogs.

---

[1] Craig Lorimer. "The presettlement forest and natural disturbance cycle of northeaster Maine." *Ecology*, Vol. 58 (1977): 147

[2] Douglas M. Griffith and Carol L. Alerich. *Forest Statistics for Maine, 1995*. Resource Bulletin NE-135. USDA Forest Service. N.E. Exp. Sta. Broomall. 1996. p. 112.

[3] New Hampshire Forest Sustainability Standards Work Team. *Good Forestry in the Granite State: Recommended Voluntary Forest Management Practices for New Hampshire.* NH Div. Of Forests and Lands, DRED, and the Society for the Protection of New Hampshire Forests. 1997. p. 103.

Tree trunks create habitat for many forms of wildlife. Some, like certain types of lichens, prefer the rough bark of very old trees. Many bird and mammal species use cavities in dead or dead-topped trees. The bigger the trunk, the more wildlife varieties that can be accommodated.[1]

Minimum tree diameters for cavity-using species	(from NH Forest Sustainability Standards Work Team, 1997)
<8”	12-18”
Black-capped chickadee Downy woodpecker* Boreal chickadee Tufted titmouse House wren Winter wren Eastern bluebird	Eastern screech-owl Three-toed woodpecker* Black-backed woodpecker* Northern flicker* Great crested flycatcher Keen’s myotis Indiana myotis
6-12”	>18”
Northern saw-whet owl Hairy woodpecker* Yellow-bellied sapsucker* Red-breasted nuthatch* White-breasted nuthatch Brown creeper Chimney swift Southern flying squirrel Northern flying squirrel ermine	Wood duck Common goldeneye Turkey vulture Barred owl Pileated woodpecker* Silver-haired bat Gray squirrel Red squirrel Porcupine Marten Fisher Long-tailed weasel
	>24”
	Little brown bat Northern long-eared bat Gray fox Black bear raccoon
*Primary excavators	

Goals. Low-impact practitioners encourage an increase in average diameter, and an increase in quality leading to the highest value forest products. They also leave some large-diameter "wildlife trees" for cavity-nesting species. They take special care not to break the bark of residual trees, especially during seasons of highest vulnerability.

Roots

Principles. The roots are where the tree takes up water and nutrients. The roots are also the anchors that prevent the wind from blowing trees over. Damage to tree roots, especially at the drip line, can degrade the value of forest products by leading to rot and decay in the stem, and can slow growth or even kill the tree.

The tree takes in nutrients through the finer roots, which are mostly in the top six inches of soil around the drip line of the tree, but can extend more than two times the distance of the tree canopy. Many varieties of fungi, called *mycorrhizae,* have adapted to extending these fine roots and increasing the uptake of nutrients and water. For this service, the trees supply carbon for the

[1] Ibid. p. 57.

fungi. Recent research indicates that these fungi can connect one tree to another--even trees of different species--and supply carbon from trees growing in sunlight to those growing in shade.[1] Severe soil disturbance, changes in soil chemistry, and lack of dead-woody material on the ground can harm these important tree allies.

The bigger the tree crown, the bigger the root system. Suppressed trees (trees that do not have dominant crowns in the canopy) often have weaker root systems. If the canopy is opened by removing the larger, dominant trees, these suppressed trees will not be windfirm and can blow over. This is one problem with *diameter-limit cuts* (where the logger cuts all trees over a certain diameter) of softwoods.

When large trees blow over, the roots pull up nutrients from lower soil layers, making them available at the surface. The uprooted large trees create a *pit and mound* structure on the forest floor that creates very different microhabitats due to differences in topography, moisture, and nutrients. Pit and mound forest floors are typical of old growth forests.

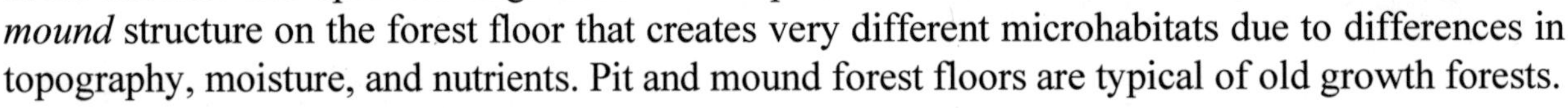

Tip-up mound from windthrow. Photo by Mitch Lansky

Goals. LIF practitioners strive to minimize root damage by minimizing impacts of heavy equipment on the soil. LIF practitioners also ensure that residual stands in vulnerable stand types are windfirm by avoiding opening the canopy too much and by leaving the dominant, windfirm trees.

Soil

Principles. The soil is the forest foundation and nutrient recycling center. The soil provides nutrients to growing plants and receives nutrients from dead plants and animals. A key area in the soil horizon is the *organic pad*, where decomposing organic matter meets the mineral soil and forms humus. This area has the most biological activity, and is where fine roots thrive.

The organic layer is built up by fallen leaves, branches and trunks. Hardwoods tend to have deeper roots than softwoods, and can pull nutrients from deeper soil layers to deposit them on the surface in the form of leaves and branches. The organic *litter* is deepest and breaks down more slowly under softwood stands. Soil under softwoods also tends to be more acidic. Organic matter is broken down by fungi, bacteria, soil invertebrates and other organisms. The organic material is

---

[1] S.W.Simard, D.A.Perry, M.D.Jones, D.D.Myroid, D.M.Durall & R.Molina. "Net transfer of carbon between ectomycorrhizal tree species in the field." *Nature*. Vo. 388/7 August, 1997, 579-582

thus not just a nutrient source, it is also an important habitat for many species. Organic matter acts as a sponge and retains water, even during dry periods. Because of this, large rotting logs can be an important reliable site for regeneration of some tree species, such as red spruce.

Disturbance of the organic pad impacts nutrient cycling and availability. Approximately half of upper soil volume is made up of pore space filled with air and water. Compaction and rutting can compress these pores between soil particles, preventing air from getting to tree roots and halting much of the biological activity. Water no longer can slowly filter through the soil, but instead forms pools or runs off, taking soil particles and nutrients with it. Soil disturbance is less of a problem when the ground is frozen or dry.

Exposing the soil surface to direct sunlight through large openings can lead to increased temperatures and more rapid breakdown of organic matter, leading to a leaching of nutrients if there is inadequate living vegetation to take it up. This leaching, combined with removal of large amounts of biomass from a heavy cut, can have a major impact on available nutrients. It can also change soil chemistry, making the soil more acidic and less fertile. Whole-tree removals have the most severe impacts since more than half of the nutrients are in the branches, tops, and leaves.[1]

An *ecological rotation* is the time it takes for the soil to recover organic matter and available nutrients. This time period depends on site and forest type as well as the intensity of cutting. The more intense the removals, the longer the recovery time. If removals and leaching occur faster than the recovery time, the soil experiences nutrient capital depletion, eventually lowering productivity.

Goals. LIF practitioners strive to cause minimal soil disturbance, and, where damage is unavoidable, to isolate it to the smallest possible area. They pay attention to timing of cut, entering stands when conditions are least vulnerable. They leave plenty of organic matter, including tops, branches, and even trunks, to rot. Where soil has been disturbed or nutrients depleted, the LIF practitioner will allow ample time for recovery.

Water

Principles. The purest water is that which comes from a relatively undisturbed forest. The closed canopy shades the soil, keeping water cooled. The canopy also protects the soil from the impact of direct rain drops and erosion. The organic matter and soil structure act as a filter to keep water clean. Forest vegetation and wetlands help to moderate extreme water level fluctuations. Forest vegetation also takes up nutrients as organic matter rots and breaks down. When there has been extensive removal of forest vegetation, some of these nutrients can leach out of the forest and pollute streams.

Water quality is most impacted by exposed, disturbed, or compacted soils where particles are apt to be washed away with the rain. This happens most often on roads, yards, trails, and stream crossings. States with forest industries all have *Best Management Practices* (BMPs) standards to reduce siltation and other water quality problems from such areas. In Maine, BMPs are voluntary.

Where water meets the forest--the *riparian zones*--some of the richest wildlife habitats are found. These areas are rich in both plants, which take advantage of increased moisture and nutrients, and animals, which appreciate the lush plant life as well as access to water. Riparian zones, if wide enough to be functional habitat, can also be corridors for migrating animals.

---

[1] James W. Hornbeck and William Kropelin. "Nutrient removal and leaching from a whole-tree harvest of northern hardwoods," *Journal of Environmental Quality,* Vol. 11, No. 2 (1982): 312

The smaller the streams, the more sensitive they are to changes in temperature, siltation, and water-level fluctuations. Opening the canopy by as little as 25% in the zones near sensitive water bodies can impact water quality.[1] Some water bodies, such as *vernal pools* (which can be breeding grounds for amphibians), intermittent streams, or small wetlands, might be wet for only part of the year, but they are important for both wildlife habitat and flood control, and the surrounding forest needs to be managed carefully.

Goals. LIF practitioners carefully control stocking and soil disturbance to maintain quality water from the forest. They pay special attention to riparian zones, especially around the most sensitive streams where these management zones should be wider, not smaller.

Wildlife habitat

Principles. When most people hear the word "wildlife," they think of moose, deer, grouse, and ducks. Occasionally they might include fish, raccoons, or even bald eagles. When biologists say "wildlife" they mean all animals (including insects and spiders), plants, and fungi. The diversity of all wildlife at all levels of organization, from genes to ecosystems and biomes, is called *biodiversity*. The diversity of wildlife helps build resistance to severe disturbances and resilience in recovery from such disturbances. Another word for this resistance and resilience is *stability*.

Forests are not just trees. They are complex ecosystems that include plants and animals. Plants create food from light, water, and minerals. When animals, such as deer, rabbits, or insects, browse intensively they can change plant distribution. Other animals, such as moths, birds, or bats, aid in plant pollination. Many types of birds and mammals help with plant seed dispersal. Some animals, such as spiders, salamanders, foxes, or coyotes, prey on others, preventing over-population and over-grazing. Other creatures, such as insects or fungi, decompose the dead plants and animals, helping them to break down into nutrients usable by plants.

Wildlife cannot survive without acceptable habitats to give them food, shelter, and other necessities. Thus the way to ensure the persistence of all species is to ensure the presence of all required habitats. The habitats must be large enough to support viable populations. A major threat to biodiversity is from habitat *fragmentation*, where the habitat is reduced so that it is too small or too isolated to support breeding, dispersal, or migration. While some species (such as deer or cowbirds) thrive on the *edge* between two habitats, others shun the edge and prefer the *interior* where there may be less predation or competition.

Fragmentation can make a habitat so small that it is all edge, and thus not be able to support interior species. In Maine, the Land Use Regulation Commission requires a minimum of 75-foot buffers (that provide “shade”) around small streams that drain more than 300 acres. When these buffers are surrounded by clearcuts, they are too narrow to support viable populations of forest interior song birds, such as bay-breasted warblers.[2] Fragmentation is most severe in populated areas where species such as cats, raccoons, or cowbirds can harm nesting interior birds.

To have all habitats requires having all the successional stages for each forest type. Due to disturbance and growth, forest habitats are always changing, both on site and over the landscape. Early-successional habitats transform over decades. Late-successional habitats can persist for centuries--though changed with small gaps. A deficit of early-successional habitat in the

[1] Steve Kahl. *A Review of the Effects of Forest Practices on Water Quality in Maine.* Water Research Institute, University of Maine, Orono. 1989: 16

[2] Brad Meikeljohn. *The value of buffer strips as riparian habitat in northern forests.* M.S. Thesis. University of Vermont. Burlington. 1994

landscape can be remedied in a matter of days with a chainsaw. It might take a century to remedy a deficit of late-succession habitat. Therefore, planning for late-successional habitat is a higher priority.

While early-successional species tend to be aggressive recolonizers of disturbed areas, late-successional species tend not to be as good at rapid dispersal and recolonization.[1] Recolonization of disturbed areas by these slower-responding species is best assured if there are species sources within (*biological legacies*, the species and habitats that survive a disturbance), adjacent to, or near the developing stand. In Maine's presettlement forest, late-successional forests made up the majority of the landscape, and there was enough time between severe disturbances to ensure recolonization and thus the survival of all species.[2]

Goals. LIF practitioners strive to provide habitats adequate to support viable populations of all native species and to assure the presence of these habitats in the landscape over time. This means having representation of all successional stages. In the Acadian forest type, late succession should be the landscape context, not a minor content. Habitats of rare or sensitive species require a higher level of protection.

III. LIF Logging Guidelines and Standards

How does one get the generalizations of the principles and goals turned into more specific practices? Who is responsible for their implementation?

Responsibility

Low-impact forestry is a partnership between the landowner, the forester, and the logger. It starts with the landowner, who has to know what LIF is and has to make the decision to do it. For LIF to happen, however, the landowners (if they do not do the work themselves) must be able to contact qualified foresters and loggers who have to agree to work within LIF standards. Otherwise the cut may not turn out as the landowners wish.

Low-impact forestry is not just a single cut--it involves long-term planning. The forester needs to incorporate LIF goals and principles into the management plan as well as ensure that loggers meet performance standards. To the extent that the forester (or logger) can locate the highest paying markets, LIF becomes more viable for all participants.

Without the logger, LIF could not happen, despite the best long-term plans. The logger must understand the techniques and have the appropriate equipment to perform LIF. Before cutting a stick of wood, the logger must know the best markets to ensure that the wood is cut to the optimal lengths considering diameter and grade.

For low-impact forestry to work, therefore, the landowners, foresters, and loggers involved must all understand the goals and principles of LIF. And they must all agree to abide by those goals and principles. And the foresters and loggers must follow basic guidelines and standards.

Forester guidelines

---

[1] Cathy Elliot, ed., 1999. Grow Flatebo, Carol Foss, and Steven Pelletier, *Biodiversity in the Forests of Maine: Guidelines for Land Management*, University of Maine Cooperative Extension Bulletin #7147: p. 103

[2] Lorimer, p. 147.

Stand Assessment. Before coming up with a management plan, the forester must assess and map the stand, taking into consideration such factors as stand types, species, volume, quality, watersheds, and wildlife habitat.

Landscape planning. Watersheds, ecosystems, wildlife ranges, and disturbance patterns do not normally coincide with property boundaries. For landowners who own thousands of acres, landscape planning starts to become possible. With smaller ownerships, the foresters and landowners should try to cooperate on a community basis to ensure that wildlife needs (such as effective corridors for migration and dispersal) are met. Cooperation of this sort, involving government, industry, and small landowners, is now being done, for example, in New Brunswick's Greater Fundy Ecosystem.[1]

Where landscape planning is possible, foresters should ensure that a representation of ecosystem types and sensitive habitats are protected in reserves. Reserves can serve many functions, from baseline "controls" for the long-term experiment of forest management, to *refugia* (source areas for future recolonization of disturbed forests). Maine currently has less than 2% of its forest in reserves. A number of organizations, such as the Forest Stewardship Council, have called for at least "representative samples of existing ecosytems."[2] In general, the bigger the landscape planning unit, the bigger the reserve system. Bigger reserves can survive the largest expected disturbances without losing essential habitats, and habitats are more apt to support viable populations of species needing larger ranges. Large reserves in the landscape, generally, are publicly owned, but small baseline reserves (to serve as an uncut control to the forestry experiments) on private land can be important as well.

Allowable cut. A number of methods can be used to calculate allowable cut. In doing these calculations, the forester must account for areas where there will be no cutting or less cutting because: 1) the site has such low productivity that sustainable management is not economically viable; 2) the site is environmentally sensitive (riparian zones, deer yards, slopes, species of special concern); or 3) the site is in a baseline reserve.

Because the degree of tree crown closure (and thus residual stocking) is so important for both productivity and wildlife, one favored method is to classify stands as "operable" when they have more than a minimum above a recommended stocking level to allow a commercial cut. The quantity above the minimum stocking is the allowable cut.[3] Another method is to ensure that cut is less than growth. Over a rolling 10-year period (for larger management units), cut might average less than 70% of growth, allowing some growth to be reinvested into the ecosystem. This calculation is normally not done at a stand or woodlot level. The area is too small and the harvest too infrequent to use 10 years as the base. Often harvests occur at intervals of 20 or more years, so a longer time frame can be used. On large ownerships with balanced age classes and

---

[1] Greater Fundy Ecosystem Research Group. *Forest Management Guidelines to Protect Biodiversity in the Fundy Model Forest.* New Brunswick Cooperative Fish and Wildlife Research Unit, University of New Brusnswick, 1997.

[2] Forest Stewardship Council Principles and Criteria for Forest Management, February, 1996, Principle #6.4

[3] Menominee Tribal Enterprises. Forest Management Plan 1983-1997, pp 1-16 in R. Simeone, M. Krones, L.Nesper, *Sustainable Management of Temperate Hardwood Forests: A Review of the Forest Management Practices of Menominee Tribal Enterprises, Inc.* Submitted to Green Cross Certification Company, Feb. 14, 1992.

logging occurring annually, the 10-year time frame can be used. Cut can, of course, exceed growth for *species* that are over-represented (such as balsam fir or red maple) and less desired for long-term stability. There are also periods of heavy mortality (such as from spruce budworm) where the cut/growth ratio over 10 years will be inadequate as a guideline for some species.

Because the majority of land should eventually be classified as sawtimber (to ensure that relatively closed-canopy mature and late-successional stands are the landscape context, not just a small content), *even-aged* management (where a cut is made that reduces the stand to seedlings and saplings, leading to a single age class) should be done only if uneven-aged management will not work for the stand. Priority for even-aged management should first go to *irregular shelterwood* (where some of the overstory is retained), and only go to regular *shelterwood* (where regeneration is well-established before cutting the overstory) if retention of residuals is not possible. *Rotation* (the interval between stand establishment and the final cut) for even-aged stands should be based on stand type and should allow enough time for soil recovery and habitat recovery, including tall, large diameter trees.

Cutting cycle. More frequent, light cutting (every 5 years, for example) creates the potential for increased residual damage. Less-frequent (every 20 or 25 years), heavier cuts create potential for more drastic stand changes. The forester can reach a compromise between these two possibilities. Low-impact logging creates an opportunity to more successfully do lighter cutting and still minimize damage on 10- to 15-year cutting cycles.

Residual stocking. The forester will consult silvicultural guides appropriate to the stand type. To ensure relatively closed canopy areas in large blocks (for adequate interior species habitat), minimum stocking should be at least 65% of crown closure, increasing to 75% of full crown closure for riparian areas. Near riparian areas, to prevent changes to water quality and flow, cuts should not exceed 25% of standing volume.[1]

Crop trees. The forester will identify crop trees and potential crop trees--trees that have good form and quality. These are the trees to leave after harvest and should be given special attention to avoid any injury that would diminish value. The normal target is around 50 to 75 per acre. Common terminology calls trees acceptable growing stock (AGS) and unacceptable growing stock (UGS). Using this approach of AGS and UGS, a harvest can be designed to improve stands and focus on the future crop trees.[2]

Pecking order. The forester should mark trees to be cut based on a "pecking order" that would prevent highgrading and thus stand degeneration. First to be cut should be high risk (trees that would not survive to the next cut), low vigor, and poor quality trees (UGS).[3] With a pecking order, the logger would be more likely to cut short-lived, poor quality medium-sized suppressed trees than long-lived, high quality, large-diameter dominant trees that are still growing well.

---

[1] Steve Kahl, p. 16

[2] Si Balch, personal communication

[3] Menominee

Mast trees. Mast trees are those that produce edible nuts, seeds, and fruits that are important for wildlife. If no high-quality (for lumber) trees are suitable for mast, some low-quality mast-producing trees (such as beech) should be retained.

Dead wood. The forester will consult recommendations from forest wildlife guides to determine a minimum of snags, dead trees, and dead-downed trees.[1] Preference will be given for larger-diameter (over 18 inches) *leave trees* (trees left behind), and allowance will be made to develop *recruitment trees* (trees that will be allowed to develop eventually into large-dead trees), since current dead-standing trees eventually fall over. The additional factor of safety must be considered since dead snags and branches have a higher potential for injuring loggers.

Logger guidelines

Felling and limbing. LIF loggers will use directional felling to avoid damaging residual trees. Limbs will be left in the woods to provide wildlife habitat and to rot and supply nutrients.

Getting trees to trails. LIF loggers will move single large stems or a few small stems (but not winch whole trees) to the trail. If winching, the logger will, if necessary, use snatch blocks to avoid damaging valuable crop trees. The logger will avoid digging up the soil during winching and use such items as grapples or cones when needed.

Wood trails. Wood trails will not exceed 10 feet wide (to give several feet clearance to machinery) unless dealing with very large trees requiring large equipment. The goal is to encourage crown closure over the trails, rather than have a series of openings large enough to allow shade-intolerant plants to proliferate. Machinery wider than 7.5 ft. should be avoided, unless trees are very large and smaller equipment will not do the job.

LIF practitioners should strive to distribute trails more than 100 feet apart to minimize damage to soil and roots. Some low-impact practitioners with radio-controlled winches distribute trails 150 feet apart. With horses yarding to forwarders, trails can be up to 300 ft. apart. With widely-spaced trails, use of larger forwarders may be appropriate.

Getting trees to yards. The LIF preference is to carry rather than drag bunches of logs. A forwarder is thus preferred over a skidder. Use of short logs, rather than tree-length logs, minimizes damage when going around curves. The more passes over a given area, the more likely the damage. So, LIF practitioners try to keep heavier equipment on permanent, widely-spaced trails to confine the worst damage to the smallest area possible.

The result, however, is more important than the method used. If a logger can use a small skidder and do minimal damage, then the skidder is acceptable. It is possible to make a single pass on a temporary trail with a narrow machine working on dry or frozen soils and do minimum damage. Whole-tree removal with a grapple skidder, especially of hardwoods, violates too many LIF principles and has the potential to cause too much residual damage to be acceptable in most cases.

---

[1] For example, *Good Forestry in the Granite State* recommends 6 trees to the acre with 1 exceeding 18 inches and 3 exceeding 12 inches in diameter. The FSC Acadian Working Group recommends 10 snag trees per acre or leaving behind the tallest tree with each cut. *Biodiversity in the Forests of Maine* has a chart (p. 146) of the snag requirements of primary cavity excavators in Maine.

Residual damage. For long-term forestry, the residual, or "crop" trees must not be damaged. During cutting, winching, and transporting trees, every attempt will be made to avoid such damage. While damage to tops and branches is of concern, it is even more important to avoid damage to trunks and roots.

Low-impact yarding area with small forwarder. Photo by Mitch Lansky

Some LIF practitioners in New England guarantee that they will damage less than 5% of crop trees. This figure is also a goal in Sweden, where any opening in the bark bigger than a matchbook is counted as "damage."[1] William Ostrofsky has developed a method for measuring damage levels for the American Pulpwood Association.[2]

Yarding. LIF yarding areas can be kept to a minimum in size with minimum damage to soil if short logs are piled with a loader, rather than pushed with the dozer blade of a skidder. LIF practitioners normally need less than 1500 square feet for yards on average. Whole-tree yarding with grapple skidders and delimbing in the yard require too much space, are too damaging to residuals and soil, and remove and damage too much organic matter to be suitable for LIF.

Truck Roads. Road width and densities should be minimized. Road rights-of-way widths should be kept between 15-30 feet with a maximum of 33 feet.[3] Generally the running surface would be 12 feet with added width if ditching, turnouts or space for snowplowing are needed. Outsloping, dips, and waterbars can reduce need for ditching.[4] Road density becomes an issue in bigger blocks of non-settled forest. Managers in the Greater Fundy Ecosystem recommended keeping road density to less than 0.9 miles per square mile due to impacts on large predators and other sensitive animal.[5] For narrow truck roads that are used infrequently, the density can be more than 2 miles per square mile.

Landscape conversion. Loggers and managers should strive to keep the percent of forest taken out for permanent trails, yards and roads to less than 15%.

---

[1] Thomas Beier, Swedish forester, personal communication

[2] American Pulpwood Association. *Professional Mechanical Harvesting Practices.* Rockville, Md. 1997, pp. 24-29

[3] Herb Hammond. *Pacific Certification Council Standards for Ecologically Responsible Timber Management.* Slocan Park, BC. 1996, p. 36.

[4] For a more thorough treatment on forest roads, see *A Landowner's Guide to Building Forest Access Roads* by Richard L. Wiest, USDA Forest Service, Northeastern Area, NA-TP-06-98, Radnor, PA, 1998. 45 pgs.

[5] Greater Fundy, p. 12

Water quality. LIF loggers will follow state BMPs to prevent soil damage that leads to siltation of waters. In addition, foresters will take into account soil type, watershed characteristics, and season of cut to further advise loggers as to when logging standards should be even stricter than BMPs.[1] Preference for LIF practitioners is to log when the soil is frozen or dry.

Conclusion

One low-impact cut does not qualify as long-term forestry--but that one cut should keep the possibility of long-term forestry open. After that cut, managers can still work with a well-stocked forest. After a heavy high-grade operation, they can't.

These guidelines are not strict rules. Following them increases the chance of meeting the goals, but common sense and experience may find other or better ways in a given situation. This document is also not the final word--the author hopes that feedback from those working on the ground will lead to improvements and refinements.

Getting low-impact forestry to work on the ground will depend on issues covered in later chapters such as: technologies, logger payment systems, contracts, assessments, and economics. Understanding the basic principles and goals of low-impact forestry is the starting place if forestry is going to act as if the future mattered.

Pesti- Side Bar:
Chemical Pesticides and Low-Impact Forestry

Aerial spraying of chemical insecticides or herbicides hardly qualifies as "low-impact." Pesticides do not always hit or stay on their target, and even when they do, they can cause ecological harm. Aerially-applied chemicals can drift or run off into ground water, causing health and environmental problems.[2] For those concerned about retaining the "little things that run the forest," broad-spectrum pesticides should be shunned.

Broad-spectrum chemical insecticides (such as those used to combat the spruce budworm) kill more than just the "target" species. They also tend to kill predators and parasites (or decimate the food supply for these species), pollinators, and aquatic invertebrates. While in some cases the affected species may rebound within a year, this is not always the case. Researchers could find no recovery in stoneflies, for example, three years after they had been sprayed with carbaryl during spruce budworm spray programs. Not all of the "target" species get killed. Given that their food supply is still intact and their predators may be reduced, "pest" populations can rebound to be a continued problem.

The guiding philosophy for dealing with "pests" in low-impact forestry is Integrated Forest Management, rather than Integrated Pest Management. The focus is on the whole forest, rather than just one element of the forest. Low-impact practitioners strive to manage the stand and landscape to be resistant to and resilient from disturbance. Part of encouraging such stability is to encourage a natural diversity of species and structure. Old-growth spruce forests, for example, have weathered numerous budworm outbreaks over the last few centuries. Part of the reason for their success is adequate habitat for predators and parasites of potential pests.

---

[1] See Janet Cormier, *Review and Discussion of Forestry BMPs*. MDEP and USEPA. 1996. And Steve Kahl's *A review of the effects...*

[2] For more detail and full documentation on impacts of forestry pesticides on the environment and human health, see *Beyond the Beauty Strip: Saving What's Left of Our Forests*, by Mitch Lansky, published by Tilbury House Publishers, 1992.

Low-impact practitioners manage away from shorter-lived, more vulnerable species toward less vulnerable species.

If ecosystems are threatened by a pest outbreak (especially exotic species that may have few natural controls), the low-impact practitioner would favor physical (removal of breeding sites, for example), cultural (managing to favor less vulnerable species and encouraging more diverse habitat), and biological approaches (such as release of predators, parasites, or diseases) over the use of broad-spectrum pesticides.

Landowners use herbicides in response to proliferation of "brush," pioneer hardwoods, and hardwood "competition," after heavy cutting. While landowners may call these unwanted plants "weeds," these weeds are often native plants that are adapted to the very habitat the landowners are creating with heavy cutting. Some companies *plan* for herbicides to follow clearcuts and plantation establishment. This is adding an insult to an injury.

Heavy cutting not only removes large volumes of wood (with all the nutrients within), it also leaves bare ground that is exposed to sunlight and the direct impacts of rain. The increased temperatures promote more rapid breakdown of residual organic matter, leading to accelerated leaching of more nutrients from the forest ecosystem. Pioneer plants slow the leaching, protect the soil with shade, and act as "nurse crops" for the more shade-tolerant species that follow. Herbicide spraying leads to: renewed nutrient leaching; damage of broad-leaved trees that have both ecological and economic value; and even, in some circumstances, some damage to the "crop" species. It is quite probable that the herbicides also affect soil microlife, either directly (through toxicity) or indirectly (through killing plants that have important interactions with soil microorganisms). To the extent that the knock-down of vulnerable plants opens up bare soil again, plants more resistant to the herbicide can fill the gaps, sometimes leading to a "need" to spray again.

Early-successional habitat (the major target of herbicides) can be avoided by avoiding unnecessary heavy cutting that creates large openings. Where overstory removal is unavoidable, it is best done in an irregular-shelterwood system, where advanced regeneration is established first and some windfirm overstory trees (or clumps of trees) are retained. Retention of some live and dead large trees creates biological legacies that help the forest recover more quickly. More targeted vegetation control, with chainsaws or clearing saws, can encourage quality trees of favored species. Having some early-successional habitat, however, is an important part of landscape management. Many species prefer such habitat for food and nesting. Wind, fire, and other natural disturbances will create such openings in the landscape. Humans are not needed to do what nature will do anyway. It is best to work with, not against, the species that are best adapted to the habitats created by such forces.

References Consulted:

Briggs, R. et al. 1996. *Assessing compliance with BMPs on harvested sites in Maine.* UM CFRU, Orono, ME.

Bureau of Public Lands. 1988. *Public Reserved Lands of Maine Integrated Resource Policy.* BPL. Dept. of Cons., Maine. 60 pp.

Cormier, Janet. 1996. *Review and Discussion of Forestry BMPs. MEDEP & USEPA.* Orono. 29 pp.

Certified Logging Professional Program. "A Guide For The Maine Logger", March 1997

Elliot, Cathy ed., 1999. Grow Flatebo, Carol Foss, and Steven Pelletier, *Biodiversity in the Forests of Maine: Guidelines for Land Management*, University of Maine Cooperative Extension Bulletin #7147

Forest Stewardship Council. 1996. *Principles and Criteria for Forest Management.*

Forest Stewardship Council, Acadian Forest Region. 1997. Certification Standards for Best Forestry Practices in the Acadian Forest Region. New Brunswick.

Greater Fundy Ecosystem Research Group. 1997. *Forest Management Guidelines to Protect Native Biodiversity in the Fundy Model Forest. New Brunswick Co-operative Fish and Wildlife Research Unit*, UNB, Fredericton. 35 pp.

Griffith, D.M., and Carol Alerich. 1966. *Forest Statistics for Maine*, 1995. Resource Bulletin NE-135. USDA Forest Service, N.E. Exp. Sta. Broomall. 134 pp.

Hammond, Herb. 1996. *Pacific Certification Council Standards for Ecologically Responsible Forest Use.* Pacific Certification Council, Slocan Park, BC. 51 pp.

Kahl, Steve. 1996. *A Review of the Effects of Forest Practices on Water Quality in Maine.* Water Research Institute, UM, Orono. 52 pp.

Lansky, Mitch. 1996. *After the Cutting is done, What's Left? An Evaluation of Forest Practices in Maine 1991-1993.* 27 pp.

Leak, W.B., D.S. Solomon, P.S. Debald. 1987. *Silvicultural Guide for Northern Hardwood Types in the Northeast (revised).* USDA Forest Service, Northeastern Forest Experiment Station Research Paper NE-603. 38 pp.

Lorimer, Craig. 1977. The presettlement forest and natural disturbance cycle of northeastern Maine. Ecology 58: 139-148.

Maine Council on Sustainable Forest Management. 1996. *Sustaining Maine's Forests: Criteria, Goals, and Benchmarks for Sustainable Forest Management.* Dept. of Conservation, Augusta, ME. 54 pp.

McEvoy, Thom. 1995. *Introduction to Forest Ecology and Silviculture.* University of Vermont Extension. Burlington. 72 pp.

New Hampshire Forest Sustainability Standards Work Team. 1997. *Recommended Voluntary Forest Management Practices for New Hampshire.* New Hampshire Division of Forests and Lands, DRED, and the Society for the Protection of New Hampshire Forests. 65 pp.

Simeone, R., M. Krones, L. Nesper. 1992. *Sustainable Management of Temperate Hardwood Forests: A Review of the Forest Management Practices of Menominee Tribal Enterprises, Inc.* Submitted to Green Cross Certification Company.

Scientific Certification Systems. 1995. *The Forest Conservation Program: Program Description and Operations Manual.* Oakland CA. 65 pgs. and appendices.

Seixas, Fernando, et al. 1996. *Forest Harvesting in the United States: A Search for Sustainable Management in Balance with the Ecosystem.* USDA Forest Service. Auburn AL. 28 pp.

Society of American Foresters. 1993. Task Force Report on Sustaining Long-term Forest Health and Productivity. SAF. Bethesda, MD. 83. Pp.

USDA Forest Service. *Land and Resource Management Plan White Mountain National Forest.* USDA Forest Service Eastern Region.

*****

Chapter 2 Endnote

How do foresters know if a forest is well stocked?

Walking through a well-stocked forest on a summer day, you immediately notice that the forest floor is more shaded and the air is cooler than in a field or a poorly-stocked forest. Foresters can tell the degree of stocking of a forest more precisely, at any time of year, by measuring the *basal area* of the stand.

When foresters talk about basal area, they are not talking about a certain location in an herb garden. The basal area of a tree is the area of the cross section of the trunk at "breast height" (4.5 feet). The basal area of a stand is the sum of the basal areas of all the trees. Depending on the average diameter of the trees and the forest type (hardwood, mixedwood, or softwood) the basal area figure can help foresters to not only determine the stand's stocking, but also (after calculating average tree height) to determine the volume.

Rather than measure and add up the basal area of every tree, foresters do random sampling, or variable plots, in a timber cruise. To measure basal area, foresters do not employ a basal thermometer. Timber cruisers usually calculate basal area by using a device called a *prism*. The prism offsets light from the trees. If the prism image intersects with the non-prism image, the cruiser counts the tree as "in." If the prism image is outside the tree boundary, the tree is "out." If the prism image is on the border, the cruiser counts every other tree (see illustration)[1].

Different prisms have different factors, which are used to calculate basal area per acre based on the number of trees that are "in.". With a ten-factor prism, for example, one takes the number of trees counted as "in" and multiplies by 10 to get the basal area per acre. If, for example, you count 60 trees on 5 plots, the average number of "in" trees per plot is 12. Multiply that by 10 and you find that the average square feet of basal area per acre is 120. With a 20-factor prism one multiplies by five.

Once the forester has the basal area per acre figure and average diameter figures (from measuring, with diameter tape or calipers, the diameter of all trees that are "in"), he or she can use stocking guides for the particular stand type to see what is the recommended residual stocking for the stand. The guides generally have an A-line, a B-line, and a C-line. The A-line shows the average stocking for stands with full crown closure. The B-line is recommended minimum stocking for adequate growth response after a thinning. The C-line is the minimum amount of acceptable growing stock for a manageable stand.[2] Below the C-line, the residual stand cannot reach full crown closure until regeneration comes up after many decades. The understocked residual stand is not making optimal use

---

[1] Illustration from Donald Sutherland, "How to measure basal area," *Atlantic Forestry Review*, Volume 8, #5.

[2] The stocking tables used in this article come from *FIBER 3.0: An Ecological Growth Model for Northeastern Forest Types*, by Dale Solomon,David Herman, and William Leak, USDA Forest Service, General Technical Report NE 204, 1995.

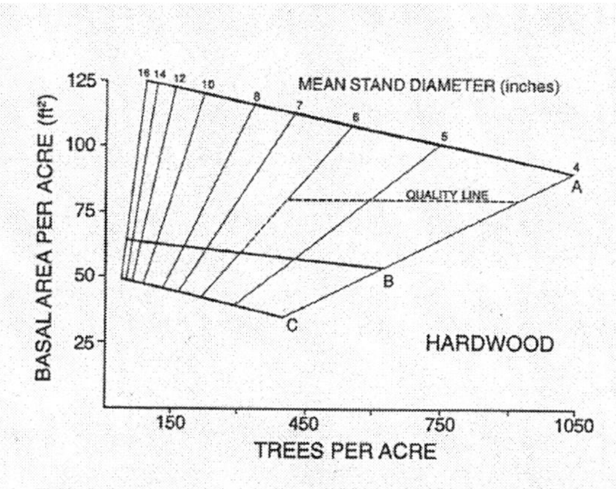

Figure 1.—Stocking chart for northern hardwoods is based on trees in the main crown canopy. The A line is average maximum stocking. The B line is recommended minimum stocking for adequate growth response per acre. The C line defines the minimum amount of acceptable growing stock for a manageable stand. The quality line defines the stocking measure in young stands for maintaining quality development.

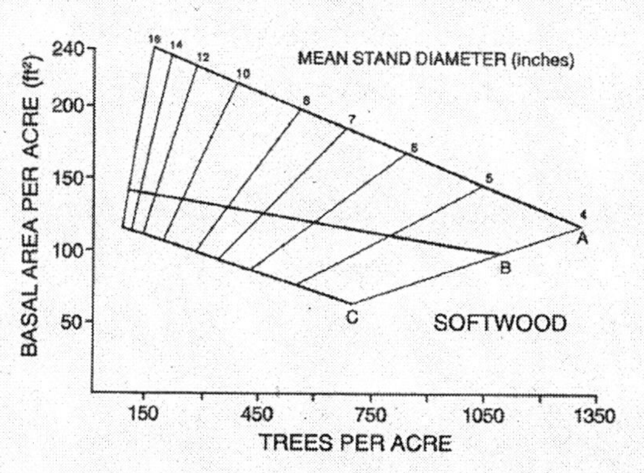

Figure 2.—Stocking chart for spruce—fir stands is based on trees in the main crown canopy. The A line is average maximum stocking. The B line is recommended minimum stocking for adequate growth response per acre. The C line defines the minimum amount of acceptable growing stock for a manageable stand.

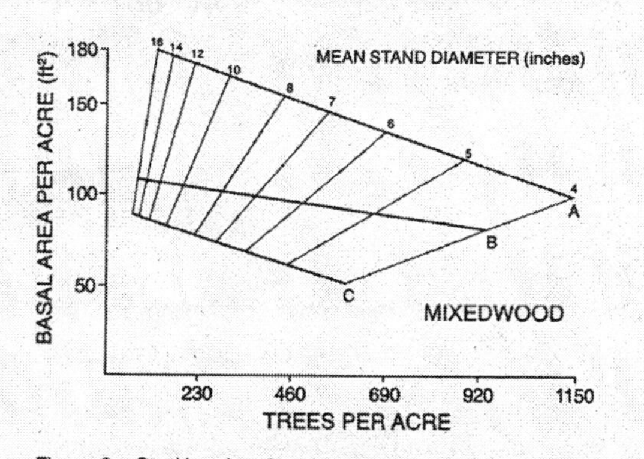

Figure 3.—Stocking chart for mixedwood stands is based on trees in the main crown canopy. The a line is average maximum stocking. The B line is recommended minimum stocking for adequate growth response per acre. The C line defines the minimum amount of acceptable growing stock for a manageable stand.

of the growing space and is not as productive as better-stocked stands. Such understocked stands are also more subject to windthrow.

The degree of stocking, and thus the amount of light coming on to the forest floor, influences the type of regeneration. Cuts that leave low residual stocking encourage a higher proportion of species adapted to disturbance and sunlight, such as poplar, pin cherries, white birch, red maple, and balsam fir. Cuts that leave higher stocking encourage domination by shade tolerant species such as hemlock, red spruce, sugar maple, and beech.

Some old-growth spruce stands, despite gaps from dead and fallen trees, have basal areas well over 200 square feet per acre. Stocking standards for buffer zones in Maine, in contrast, are quite low. The Maine Forest Service, for example, introduced legislation in 2002 to set a minimum stocking of 60 square feet of basal area in riparian zones, regardless of stand type. This is below the C-line for softwoods and mixedwoods.

Typical cutting in Maine leaves stocking well below recommended levels, especially in softwoods. An analysis of Maine Forest Service data from surveys of cutting sites 1991-1993,[1] for example showed that only 8% of all partially-cut acres had residual stands above the B-line (based on an average diameter of 8 inches). None of these adequately-stocked stands were in softwoods. In contrast, 43% of all partially-cut acres had residual stands that were *below* the C-line (based on an average diameter of 6 inches).[2] This low residual stocking is the result of heavy cutting. Half the partially-cut acres had removal rates of more than 40% of the original basal area. Not surprisingly, there has been a gradual shift in Maine towards more disturbance-adapted species (see endnote for chapter 1).

---

[1] *After the Cutting is Done, What's Left? An Evaluation of Forest Practices in Maine 1991-1993*, by Mitch Lansky, March, 1996

[2] This is a rather low hurdle, as sawmills prefer softwoods that are over 9 inches and hardwoods over 11 inches for sawlogs.

3. Ecosystem Management[1]

Low-impact forestry is a type of ecosystem management. In the following excerpt of an interview from 1997, Barbara Alexander, who helped organize a low-impact forestry program in northern Vermont, interviews David Perry about ecosystem management. Dr. Perry is professor emeritus of ecosystem studies in the Department of Forest Science at Oregon State University. He is an internationally known expert on the structure and function of forest ecosystems and landscapes, the role of biodiversity in ecosystem stability, and the interactions among ecological scales. He has authored or co-authored over 75 refereed journal articles and 13 book chapters and symposium proceedings. He has also authored or co-edited four books, including the outstanding 1994 publication Forest Ecosystems. *Dr. Perry currently resides in Hawaii, where he is involved with native ecosystem restoration and developing sustainable communities. He has taught courses at the Ranger School in Fredericton, New Brunswick, and has some familiarity with the Acadian forest ecosystem.*

BA: Ecosystem management is driven by explicit goals which differ from those of stand-level management. Would you outline or summarize these goals?

DP: Of the goals that you see most often mentioned for ecosystem management, the one that tops the list is always sustainability. Society, through ecosystem management, is coming to grips with the issue of sustainability. Sustainability by itself doesn't tell you much, so the next thing you have to ask is: sustaining what? And then the goals get more specific.

What ecosystem management strives to sustain are habitats for all indigenous species and the health of the overall system--that is, all the processes that come together to confer on the system the capacity for self-renewal. One word that has been used to describe this is "integrity." Ecosystem management strives to maintain the integrity of the forests, "integrity" meaning all the indigenous species, all the processes operating within proper bounds.

One thing ecologists have discovered over the last 15 years or so is that, if you're interested in the dynamics of an ecosystem, you can no longer focus only on a given piece of ground. Every stand of trees, for example, is part of a larger landscape, and the landscapes are part of regions, and regions are part of a globe, and all of these things come together to influence what goes on on any one piece of ground, and sometimes influence it very significantly.

Acid rain is an example of that, global climate change is another. Another large-scale concern is habitat for wide ranging species. In western North America, the spotted owl is the exemplar, but there are many others. In the Northeast, pine martens are good examples of species that, if you're going to maintain healthy populations, you have to think in terms of areas much larger than foresters are accustomed to think about.

So ecosystem management has made explicit the goal of dealing not only with stands, but with landscapes and regions. That brings on, then, the importance of cross ownership, cross boundary kinds of interactions. And so another objective of ecosystem management is to bring together multiple ownerships within an area of interest and to strategize together in order to achieve common goals.

---

[1] First published in Northern Forest Forum, Vo. 6, No. 2, 1997

BA: What you've been referring to is how ecosystem processes operate over a broad range of temporal and spatial scales. As we move from stand-level to landscape-level management, we also move from the short term to the long term in forest time--which means thinking not only in a time frame of decades, but in a time frame of centuries. What changes in current management practices need to be made to manage on the landscape and century scales?

DP: Managing on the scale of landscapes means that in order to meet landscape objectives you have to ask, "What are the landscape level processes we need to be concerned about maintaining." There are three in particular that come to mind. One is habitat. Habitats for most species must be maintained on landscapes, so if we're looking at maintaining a habitat for pine marten, then we need to think on a scale of thousands of acres.[1]

The second is the spread of disturbances [...] like budworms, fire, or pathogens. Anything that moves through space is affected by the structure of that space. So the structure of the forest has a great deal to do with the ability of disturbances to move through it.

One of the things we've seen in the Pacific Northwest--and the same, I'm sure, is true in some areas of the Northeast--is that many of the natural forests were pretty resistant to crown fires, but have been converted to forests that are highly susceptible to crown fire.

The third landscape level issue is hydrology. When you think about hydrology, the movement of water, then you must think in terms of landscapes. And again, the pattern of forest types that you put out on the landscape will influence hydrology.

In terms of time and stability over time, the basic concept that has emerged over the last ten years is that of biological legacies. Now let me emphasize--when I use the word stability, I don't mean no change. The idea that forests are stable in the sense that they don't ever change is just not true. All forests change to one degree or another, and all forests are disturbed to one level or another. Stability boils down to retaining the capacity for self renewal, which is where biological legacies come in. "Biological legacy" is a term Jerry Franklin applied to components of the system that survive disturbance and act as foci for the recovery of the rest of the system.

A good example is shrubs and trees that can sprout from roots after above ground parts are killed. These send up new shoots, which soak up nutrients and keep them from being leached to streams. The reborn plant stabilizes soils and pumps photosynthates to the myriad soil organisms that require plant carbon to survive. Legacy plants become centers of biological activity. Birds land in their branches and defecate seeds.

Pioneering work by some of my former students--Mike Amaranthus, Sue Boudreau, Suzanne Simark--has shown that newly established seedlings find a rich biological and chemical environment in the soils around these legacy plants, and that interconnections form via mycorrhizal hyphae,[2] through which carbon and nutrients flow from one plant to another.

Another example of a biological legacy that is especially relevant to forestry is big dead wood. Big dead wood has multiple important ecological roles, including acting as biological legacies, which allows certain species of invertebrates and microbes to survive fire.

---

[1] Editor's note: In managing for habitat, the manager must make sure that the habitats are large enough for viable populations, and that account is taken of access, dispersal, and migration. New habitats might not get colonized by the appropriate species if the habitats are too isolated from source populations.

[2] Mycorrhizal hyphae are fungal threads that form extensions of tree roots and increase the intake of moisture and nutrients.

We used to say that a clearcut mimics a natural disturbance. A clearcut in no way mimics a natural disturbance other than the fact that it opens up the canopy and lets light down to the forest floor. A clearcut takes away a lot of wood, it takes away many of the legacies that would have persisted through a natural disturbance.

So now, if we look at this issue from a scientific standpoint, in order to give good guidance to management, one of the central concerns we're facing is, O.K., we need to learn more about this biological legacy of big, dead wood, and exactly what it does, and exactly how important it is to the recovery of the system, and how much we need to leave out there in order to not disrupt the capacity of the system to renew itself. Experiments to clarify these questions are going on now, but it's still early in the ball game.

My own feeling is that you can take wood off of a system without disrupting its capacity for self-renewal, but you also need to leave some out there or you will eventually disrupt that capacity. So, the question now is, "How much?" And that's what we have to learn. But it's going to be a long time before experiments yield definitive knowledge, not only with regard to the function of big dead wood, but for many of the questions surrounding maintenance of ecosystem integrity. In the meantime, best judgments must be made.

BA: The concept of dead wood being the life of the forest...

DP: Actually, that's quite true, and a very good way to put it. Studies have been done looking at changes in nutrient composition as logs decay. Consider two of the most essential nutrients, nitrogen and phosphorus. In the bole of a living tree, the ratio between nitrogen and phosphorus is about 80 to 1. In an old decaying log that's been down on the ground for about 120 years, the ratio of nitrogen to phosphorus is about 25 to 1, which is very close to that of a living cell.

What that says to me is that there are far more living cells in an old decaying log that's been down on the ground for about 120 years than there are in the bole of a living tree. When you look at an old log, at least out in the western U.S. or western Canada, that's easy to understand because they're great water reservoirs. The western systems tend to have drought summers, and so where the water is, is where the life is.

BA: Prior efforts to preserve biological diversity have focused on a relatively small number of species, subspecies, and populations of plants and megafauna, yet the lesser organisms like bacteria, invertebrates, and fungi make up about 90% of the total of all species and carry out critical ecosystem functions. Will these organisms be conserved for the long term only if ecosystems are conserved?

DP: A very good question, and one that's impossible to answer precisely, because we simply don't have the knowledge. But there is no doubt in my mind that if we do not manage ecosystems, we're going to lose some of these organisms. What that translates into in terms of ecosystem function is another question. I think there's a reasonable chance that over the long run, if we erode the diversity of these little critters--E.O. Wilson called them the "little things that run the world"--eventually we erode the resilience of the ecosystems that we're managing.

Now nature is very robust, and there are presumably a lot of backup safety systems; however we're really getting into questions of how ecosystems are wired up. How do they work? What do each of these little species out there contribute? How much overlap is there between them in their function?

These are questions of redundancy versus keystoneness, something that ecologists debate fairly hotly. Some believe there is plenty of redundancy in nature, others believe that there is no redundancy at all, that everything does a unique job. I come somewhere in between those two. I believe there is redundancy, but it is not limitless. And so, as we erode diversity, we set ourselves up for what several scientists have very aptly called "unpleasant surprises."

It's like having a six-legged stool. You pull one leg off and the stool stands up and you say "Huh, that's interesting." And you pull another leg off and the stool stands up and you say, "Oh well, legs don't make any difference to the stool standing up." But that's not a valid conclusion--something we discover to our surprise when the stool falls down. With regard to the erosion of diversity and long-term stability of ecosystems, scientists don't know how many legs it takes to hold the stool up, and there's a good chance we will never know. Aldo Leopold had it right when he said "keep all the pieces."

A lot of evidence has accumulated over the years that more complexly structured systems support a greater diversity of both big and little species. The work that Tim Showalters of Oregon State University has done both in the eastern and western U.S. shows that old-growth forests tend to support a much greater diversity of spiders in their canopies than young plantations. Old-growth forests also have a much more favorable ratio of predatory insects to tree-eating insects than young plantations. George Carrol and his students at the University of Oregon find that older trees have a significantly greater diversity of foliar endophytes[1] than young trees, and young trees near their elders have a greater diversity than young trees distant from elders, another kind of legacy effect.

The evidence from these and many other studies points toward prudent behavior if our objective as a society is sustainability. Prudent behavior demands that we don't lose any species, small or large, which translates into protecting their habitats within managed forest landscapes. Even when we focus on the little things, we are led back to landscapes and regions. I often begin talks with two slides--the first of the earth from space and the second an electron micrograph of fungal hyphae within a soil aggregate. The message is that, if we want to sustain our life support systems, our thinking must stretch from the very large to the very small.

BA: So we move from the landscape scale, to the micro scale, to the history scale. I'd like to ask you a question about presettlement forests. The regional ecosystems of our northeastern forests have been forever altered because of industrialization and excessive timber harvesting. How close to presettlement forest conditions could we come using a combination of ecosystem and non-management systems? And should we try?

DP: Yes, we should move the forest back toward presettlement conditions, with the emphasis on "toward." Whether we could recreate exactly what was here before, or would even want to, is another question. The Acadian forests particularly are very different now than they were 100 years ago, 200 years ago. They are much less stable, if you talk about stability in terms of their ability to resist insects and other stresses, and they are much less diverse. It may be impossible to ever go back to what they were exactly.

---

[1] Foliar endophytes are microfungi that live symbiotically in plant leaves and help protect their hosts against pests and pathogens.

But I don't think that should be a concern. What we should be doing is diversifying the species again--reintroducing some of those that were once very important in the system and have dropped out to a large degree.[...] My feeling is that management can be an aid to this. If the goal is clearly envisioned, the management can be designed that helps attain that goal. This is not to say that every piece of ground should be managed. It shouldn't. But it is to say that management done right, with a clear vision of where society wants to go and an understanding of basic ecological principles, can be a positive thing.

BA: As our knowledge base changes, we will need to make short-term decisions for long-term planning or goals. Would you comment on how ecosystem management represents adaptive management?

DP: As I understand it, there are two central ideas to adaptive management. One is that you maintain the options to adapt, and maintaining the options to adapt means, essentially, that you maintain diversity. The second idea is that, again, you have a clear vision of where you're headed and what you need to achieve on the landscape so you can come back and look and see if what's been done is taking you in the direction you want to go. So adaptive management is retaining the ability to adapt, and collecting the information that tells you when adaptation is necessary.

Large tree marked with "W" for wildlife leave-tree at Baxter State Park Scientific Management Area. Photo by Mitch Lansky

All this is easily said; it's more tricky to do. You can argue that once you have cut a single tree out of the forest, you have, to a certain degree, foreclosed an option for the future. There's truth to that argument, but it's also a plain fact that we're going to be taking products out of many of these forests. What we need to do is figure out ways to take those products out without closing the door on being able to adapt and do things a different way.

My own feeling is that once you clearcut a forest, you've lost options to have big trees out there again for 50 years, 100 years, 150 years, or 2000 years, depending on where you are, and so that really has closed the door on that part of the forest where you have cut away those big trees. Which is not to say you can't cut big trees, but if you cut some and leave some, then you have left options out there for yourself.

BA: We've talked a bit about limits to our understanding--ecosystems have been described as moving targets with futures that are uncertain and unpredictable. So, ecosystem management must also be experimental to some degree?

DP: Oh yes--every piece of management is experimental to some degree in the sense that we really have a limited predictability about where nature may take the things that we try. And so we're mucking around and we've been doing that for the last 100 years in forestry. What we are now looking at in ecosystem management and adaptive management is formalizing that experimental aspect more so that through our experimentation, through our trying things, we can come back and look and we can learn from it, and from what we learn we can improve what we do.

My own feeling is that experimenting with different management approaches is a good thing--it's how we learn. But prudent behavior dictates the need for guidelines; filling landscapes with an experiment that fails would not be good. One rule of thumb I favor is the more a given management approach departs from the natural forest structure, the less area it should occupy (at least until its stability is established, which could take decades or centuries). This is quite the opposite of what forestry has been doing during the 20th century, which is widespread conversion of whole regions to forests that differ in fundamental ways from the natural. Of course, forestry during the 20th century was designed to meet society's demand for wood, and if society now demands other values from forests, the appetite for wood is going to have to be controlled.

BA: By what criteria will the effectiveness of ecosystem management be judged?

DP: Three criteria: It will be judged by its success biologically and ecologically--that is, its success in maintaining habitat and maintaining the integrity of systems. It will be judged by social acceptance. And it will be judged economically. And, you know, the bottom line in our economic system is that somebody's got to pay for it. On public lands, the people at least have the option to pay for part of the cost of managing ecosystems, rather than just managing trees. There's a lot of precedent for that. For a lot of years the people subsidized building roads into western forests to cut them down; now on public lands we should be thinking of subsidizing the maintenance of biodiversity and system resilience.

On private lands, it's another issue. There are private landowners who are not all that interested in how much money they can make. They want to make some money, and they want to maintain, to the extent they can, a forest that's aesthetically pleasing. Large industrial landowners march to a different drummer, obviously, and ecosystem management is going to be judged on its ability to pay for itself on these large industrial lands.

One of the things that we have learned in ecology, and in landscape ecology in particular, that translates directly into ecosystem management, is that not every piece of ground has to do the same thing in order to achieve our goals of preserving diversity and resilience and so forth. Because it isn't necessary that the same thing be done on each piece of ground, there's some leeway for one kind of management in one place, so long as the types of management and the places in which that management are done are chosen intelligently. And, by that, I mean bringing all the knowledge that we have to bear on answering the question, "Does this achieve our goals or not."

In areas like the western U.S. where there is a lot of public land which, to a large degree, carries the water for species preservation, pressure is released from the private lands. Private lands still have responsibilities in maintaining forest integrity, there's no question about that, but for many species, the burden of carrying habitat probably can be done on public land. If you go to a place like the Northeast, where there are many fewer public lands than in the West, it becomes much more difficult, in theory at least, to balance the issue of species preservation and maintaining system integrity with economics. But the fact that it's difficult doesn't mean it can't be done. Already there are certain market tools swinging into place to help that happen, especially green certification.

So it truly is a biological, social, and economic package that we're dealing with, and all those things are going to have to work together. In order to choose the balance, for example, between economics and biology, we really need to be clear about our goals. We need to be clear about the thresholds in the system and how far they can be pushed, and we need to get comfortable with the idea that sustainability requires leaving and giving back as well as taking. And we have to be very clear about this business of uncertainty, something that is seldom if ever translated into the economics of natural resource management.

Everyone who can afford it goes out and buys some insurance--this is a hedge against uncertainty. We don't think twice about why we pull some money out of our pocket to buy insurance; it's obvious to us why we're doing that. We need to translate that concept into managing natural ecosystems. It may cost something, in terms of short-term profits, in order to maintain more diversity in the system and help buffer it against disturbances and unexpected surprises. But that cost is a legitimate cost of insurance; it's the only insurance that we can buy in forestry--maintaining complexity, drawing on the mechanisms that have evolved in nature to maintain integrity, preserving those mechanisms, enhancing and restoring them where necessary. We should count that as part of the legitimate cost of doing business.

*****

4. Wood Harvesting Technologies and Low-Impact Logging

By Sam Brown

A forester makes a management plan for a forest stand, but this plan does not become a reality until a logger follows it. What type of machinery and what type of logging systems are most appropriate to achieve low-impact forestry goals? Sam Brown explores these questions in this chapter.

Introduction

LIL (Low-Impact Logging) is any system of logging which reduces known harmful impacts to soil, water and plant life, retains a functional forest after harvesting, and is economically viable for the operator. It is not a specific method of logging, but rather an awareness of the consequences of today's actions on tomorrow's forest values.

Economic Viability

LIL is part of a more complete way of looking at forestry economics. It considers not only the importance of harvested trees but residual trees as well. It looks at present and future values. It acknowledges both market worth and intrinsic ecosystem benefits (biological, social, and effects over time). *In this regard, any logging damage represents a cost to the landowner, reducing the forest's future economic and ecological value.* Therefore, LIL seeks to match harvesting technology to desired forestry outcome, and does not change the forestry to suit the machinery.

Choosing a System

Minimizing damage is the main characteristic of any LIL technology. Many current technologies can be adapted to LIL. However, the attitude of the logger is at least as important as the type of system used to remove forest products from the woods. Possessing and using "low impact" equipment is not enough; the logger must understand *why* the machinery and techniques are desirable in the first place. Good technology can cause big damage when used thoughtlessly.

Overall Goal

The ideal LIL goal in selecting what to harvest is to leave a well-stocked stand with a full range of tree sizes and nearly-full crown closure, to minimize nutrient loss, and minimize damage to soil, animal habitat, and residual trees. When considering a logging operation, seven basic environmental factors affect what technology to use:

1) *Size*--amount of land and volume of wood.
2) *Stand type*--size and kind of trees, density, etc.
3) *Slope*--steepness of the topography.
4) *Soil*--harvesting conditions (soft, wet, dry, etc.)
5) *Surface*--obstacles on the forest floor (boulders, ledge, streams, etc.)
6) *Distance*--to yard from trees, to market from yard, roads, etc.
7) *Season*--part of the calendar year when harvesting will occur (affecting soils, animal habitat, sap flow, etc.)

Choosing Machinery

Logging is getting trees out of the woods and as such has four basic operations:

- felling, bucking, and limbing single trees;
- gathering single trees into a bunch ("prebunching");
- moving bunches to the yard ("skidding", "yarding" or "forwarding"); and
- sorting and piling trees at the yard.

The trees can be moved "whole tree" (trunk, branches & all), “tree length” (with tops and branches removed), or cut up into "shortwood" (smaller pieces). Three general methods are currently used in the Maine woods for harvesting trees:

1) the conventional "manual" system (both tree length and shortwood), which consists of a chain saw and some sort of tractor to drag or carry wood;

2) the mechanized Feller/buncher (FB) whole tree system, which consists of a felling and bunching machine, a grapple skidder, a delimber and a loader; and

3) the Cut To Length (CTL) short wood system, which consists of a harvester and a forwarder.

The manual systems have the least equipment and therefore cost the least of the three (from 20 to 100 thousand dollars). FB and CTL both require large capital outlays to afford the expensive machinery and its support systems (from 250 to 500 thousand dollars).

Let's examine how these three approaches satisfy LIL requirements:

Felling, Limbing, and Bucking.

Directional felling is critically important for LIL to reduce damage to the remaining stand from the falling tree as well as positioning it to be removed from the forest with minimum damage to soil and residual trees. Leaving the limbs scattered across the forest floor is better for nutrient distribution than concentrating them in piles on the trailside or in the yard. "Bucking" is cutting the tree stem into market lengths, and is a significant step in getting the most value from a tree (for instance, when a log is cut too short for the specifications of a mill, it will either be scaled back to the next shorter length or demoted to pulpwood).

The most traditional method is manual felling, limbing, and bucking with a chainsaw. The operator in this system walks to the tree, chooses a felling direction, drops the tree, removes the limbs with the saw and, if desired, bucks the stem to whatever lengths are required. Manual directional felling requires high levels of skill and special tools to overcome adversities like lean, slope, wind, etc. It is the least mechanically complicated and the slowest of the three methods.

The cutting part of the FB (feller/buncher) method consists of a tractor which carries a felling-and-bunching head. Some FBs are wheeled tractors with the head mounted in front on a short boom, requiring the operator to drive to each tree to be cut. Others are similar to tracked or wheeled excavators, with the head at the end of a long (up to 30 feet) moveable boom, allowing the operator to collect more trees with less driving. The hydraulically controlled boom of the head has a set of jaws with which the operator grasps the tree, and a saw head (or, less commonly, hydraulic shears) which severs the tree from its roots. Limbing and bucking are not done in the woods with this system. This system has medium mechanical complexity and is the fastest producer of the three methods.

CTL is similar to FB in that a tractor ("harvester") carries a head which can fell a tree, but also can limb and buck it. The harvester's boom is usually quite long (up to 35 feet), allowing the operator to reach many trees from one location. The head grasps the tree, cuts it with a saw, directionally fells it, and moves it to the trailside, as can the FB method. But in addition, the CTL head can simultaneously limb and buck the stem. The head pulls the stem through knives which remove the limbs while measuring the length of the delimbed tree, then bucks it at the appropriate place. Often the limbing is done directly in front of the harvester's path, creating a mat of limbs over which machines can travel. This mat distributes the weight of the machine and reduces contact between the wheels and soil.

Some forwarders run on tracks which increase flotation but also are more damaging when turning, as they scuff the soil rather than roll over it (tracks also cut into duff layers and can thus potentially cause root shear). CTL is the most mechanically complex of the three systems and has medium productivity.

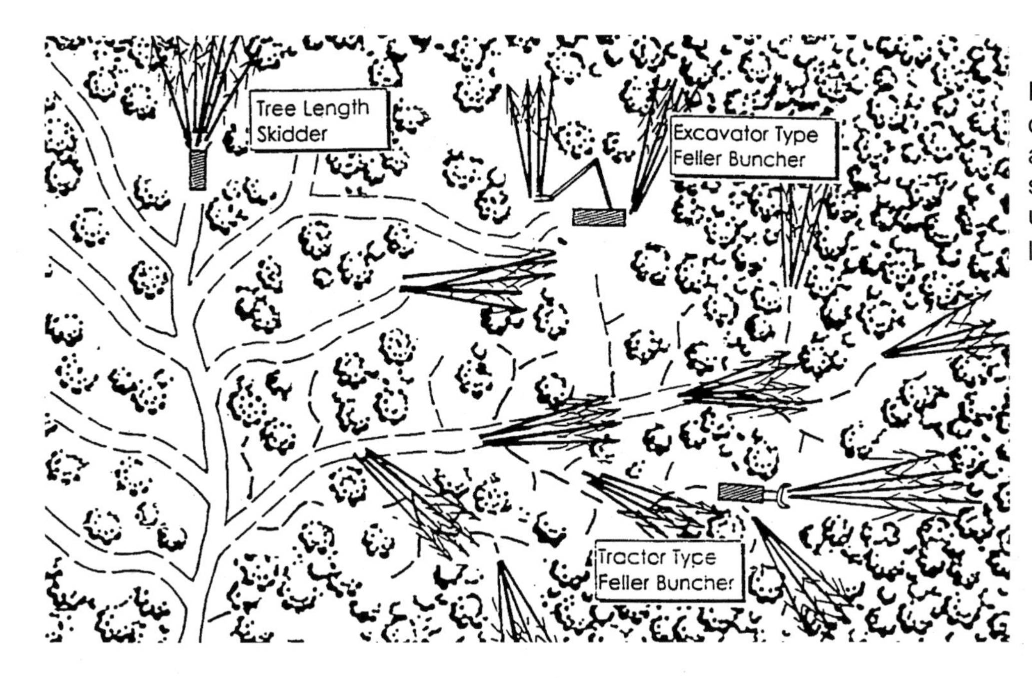

Figure 1. Diagram of typical skidder and feller buncher systems. These use 20 to 30% of land area for Trails.

Gathering single trees into a bunch

Efficiency of yarding is increased when wood is gathered into groups before yarding. Moving short pieces of trees through the woods to a yarding machine is less damaging to the soil and to standing trees than moving the machine to each piece. The gathering process consumes time and energy, but it can dramatically increase the distance between harvesting trails, which in turn keeps more land in production.

In manual methods, after cutting, limbing, and bucking, the trees are either pulled by a winch or an animal to a gathering area or directly picked up by the yarding device. Winching allows the machine to remain in one place and gather the pieces to it, significantly reducing soil compaction in the stand. Pulling single stems requires only a narrow path through the remaining trees; however moving one stem at a time is slow. Using a radio controlled winch increases the efficiency of the operator by reducing the amount of walking, and also allows careful observation of the log during winching, to avoid damage to soil and residuals and reduce interference with obstacles. Snatch blocks and conical grapples can be used to get around obstacles and further reduce soil damage. This method is the slowest of the three.

In FB systems, after the tree is cut, the operator can then tilt the head to lay the cut tree down in a specific place or proceed to the next tree to be cut, still carrying the first tree, and continuing in this manner to gather trees until the jaws are full. The gathered bunched trees are then laid

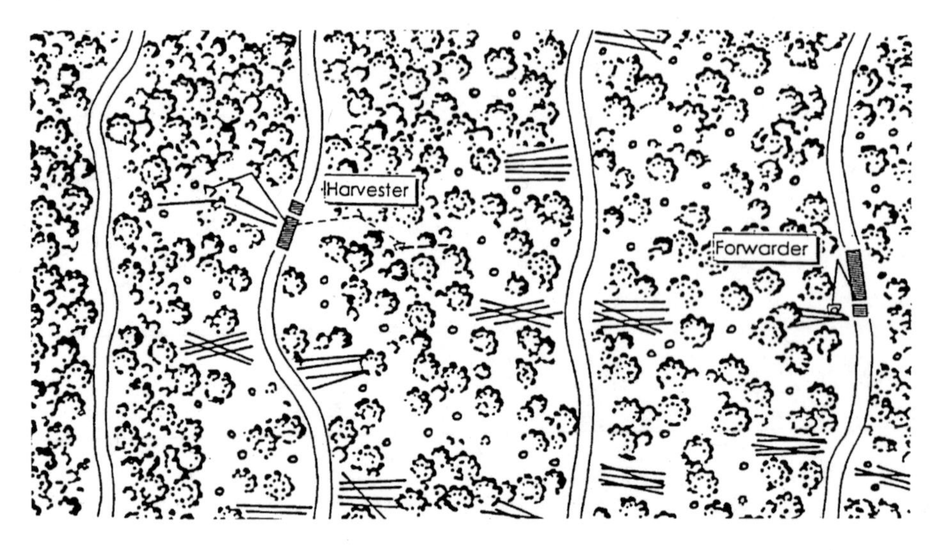

down for subsequent dragging out to the yard. Because the FB cuts and carries trees, its overall weight increases and consequently the pressure on the soil to support that weight also increases.

Figure 2. Diagram of typical Cut To Length trail system; this arrangement uses 20 to 28% of land for trails.

The CTL system is ideal in that it can carry the cut tree off the ground to the trail and thus totally eliminate soil damage between the trails and, with a careful operator, residual damage as well. The boom length of the loader determines the spacing of the trails, as it is the limiting factor in reaching into the stands (trails are usually between 30 and 60 feet apart in this system).

In CTL, while processing the tree, the operator controls where the cut-to-length pieces are placed, piling them into stacks for ease of subsequent handling. These piles can be close to the road or can be at the far end of the harvester's boom, to create a "ghost trail" (see Figure 3). Such a system effectively doubles the trail spacing distance by allowing the forwarder (having a similar length of boom) to reach into the stand from every other trail. However, piling at the end of the boom increases the likelihood of damage to the stand because the operator can't see as well. Carrying the same amount of wood out on fewer trails concentrates the damage potential for those trails.

The wheeled feller buncher is the least desirable system, because it requires moving the machine over a wide area of forest floor, compacting and rutting as it goes and damaging other trees. The CTL, because it handles limbed trees, has a maneuvering advantage over the FB excavator-based versions, which must have enough room and strength to manipulate whole trees.

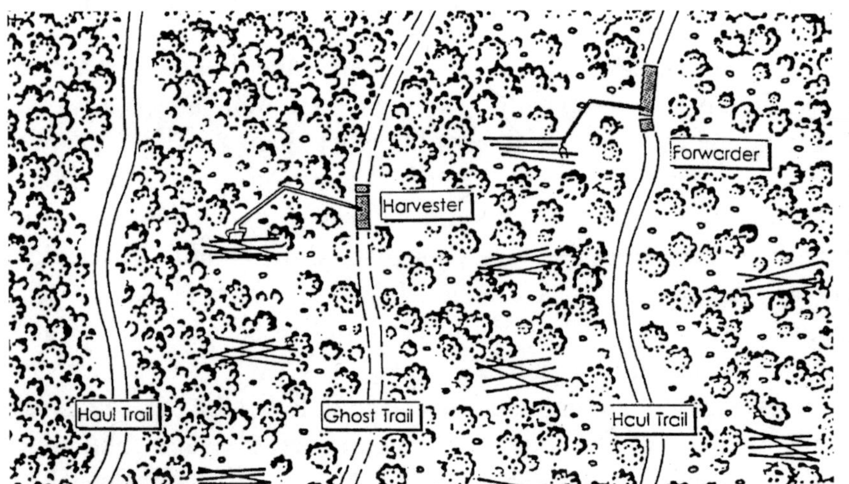

Figure 3 Diagram of CTL ghost road system. The harvester moves wood from ghost trail toward haul trail so that forwarder operates on every other road.

Manual winching or twitching offers the greatest flexibility in trail spacing because of the long length of the winch cable or the unrestricted mobility of a draft animal. Trails spaced up to 150 feet apart are not uncommon.

Figure 4. Diagram of road system for prebunching with winch or draft animal. This arrangement uses between 8 and 15% of the land for trails.

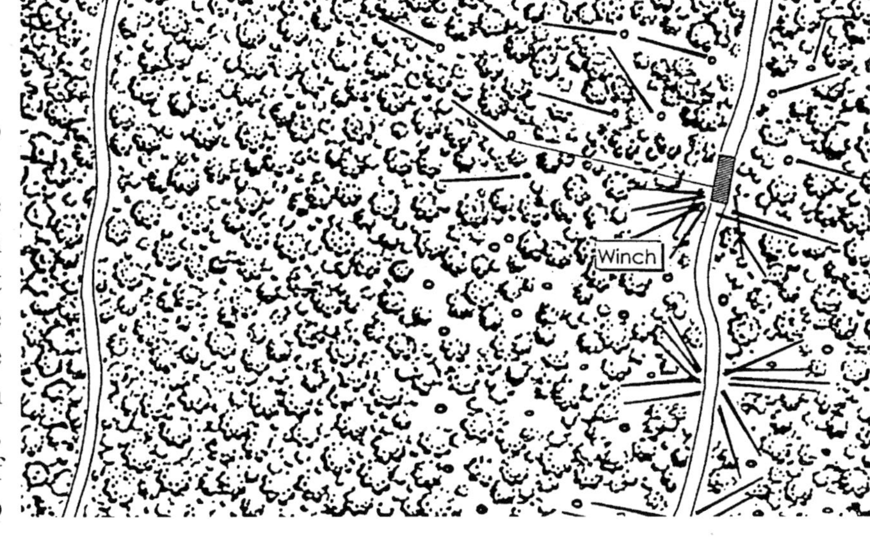

Moving Bunches to Yard
The bunched pieces by the trailside can be handled in two basic ways: short wood or tree length. The most elemental technique is to drag out tree length bunches with a skidder, tractor, or team of animals. Skidders back up to the cut trees and hook them up with wire cable or grasp them with a grapple. The FB system uses a grapple skidder to move the pre-bunched piles. Some whole tree (FB) trails can be as much as 30 feet wide (from uncut limbs spreading out) when the skidder drags the load to the yard. Some FB systems use a cross between the two, half carrying and half dragging whole tree loads. These are very fast ways to move wood, but very damaging to the road surface soils.

Alternatively, bunched wood is cut into market lengths (usually no longer than 16 feet) and loaded onto a sled or wagon and then hauled out ("forwarding"). Carrying wood on a trailer has many advantages for LIL work: short length and width and good maneuverability of the overall machine reduce the potential for damage to roadside trees; trail width is narrow; fuel efficiency is much greater for carrying trees than for dragging them; and rolling wheels do not damage the soil as much as dragged trees.

Tracked vehicles exert less pressure on the soil than equally-heavy wheeled machines, are more expensive to purchase, require more maintenance and are slower in general. Brush piled and matted in roadways reduces ground pressure on the soil from any machine.

Yarding. An ideal LIL yard would be small in acreage, have hard dry soil with no siltation into the watershed, would cause minimal damage to logs and other products from mud and machinery contact, would make sorting into many products easy, and would not be an eyesore.
In FB methods, the unlimbed trees are delivered to the yard close to a delimbing machine, which strips off the limbs and then bucks and piles the stems. Often the pile of limbs becomes a hindrance to yard activities as well as being unsightly, and in this instance is not returning nutrients to the stand.

To separate various products from their load in the yard, skidders must stop, drop the load, remove the cable from the selected product, move ahead to drop the product off and repeat the process for every type of product. Then the skidder returns to push the sorted products up into their respective piles. Turning and pushing the sorted products up requires traction from the skidder wheels, which increases the yard's soil disturbance.

Here again, forwarders have an advantage over skidders. Having the product already cut to market length, forwarders can easily pile into many sorts and can make neat high piles to minimize area taken by wood waiting to be trucked away. Such piles are more easily measurable than tree length piles.

Trucking

A significant amount of time is spent by the logger in arranging for and handling the loading of trucks. Flexibility of schedules is dramatically increased when the trucker can come and go at will and not require auxiliary traction from the logger's machinery.

When a skidder drops its load, turns and pushes it into piles in the roadway of the yard, the soil gets dug up. Loaded trucks frequently get stuck in the slippery mud of skidder-churned yard soil and require further assistance from the skidder to haul them out to more solid road surfaces. Mud falling off such a truck onto the public paved road is a hazard for motorists as well as an unattractive advertisement of the local logging operation.

Forwarders do not have to turn around much in a yard, moving straight back and forth along the piles. Soil is therefore less apt to get disturbed than with skidders.

Conclusions

LIL is result oriented. The main focus of LIL is to reduce logging damage to soil and to residual trees to the lowest practical level. Damage to roots, bark, and branches of residual trees is permanent and negatively affects the trees' future growth and value. The whole forest system needs to be evaluated when considering the impact of any specific logging operation. Biological, social, and economic factors cannot be isolated from each other.

The current competitive woods products market is based on production. The small-volume producer is at a disadvantage to a large-volume producer. High interest rates on equipment loans and low market prices for wood can force the purchaser of machinery to place volume of production above quality of work. A high cost system must produce higher volumes than a lower cost one to handle the higher interest payments. The initial purchase price of any system therefore becomes very significant. New methods of paying the producer must be developed for LIL, based on the improved future value of the landowner's assets rather than solely the current market value of the removed products.

With all this in mind, of the three harvesting methods examined, short wood harvesting and forwarding seem to offer the best combination of features to accomplish LIL objectives while maintaining reasonable levels of productivity for the operator.

*****

Harvesting equipment

Directional felling with a chainsaw

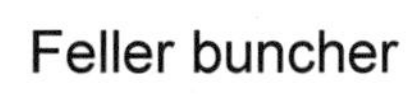

Feller buncher

Single-grip harvester/processor

Yarding equipment

Mini-skidder

Grapple skidder

Forwarder

Tools for low-impact yarding

Radio control for cable winch

Self-releasing snatch block for winching around corners

Grapple for skidding out single stems without gouging into the ground or getting hung up on stumps

Sam Brown's Low-Impact Forestry[1]

Sam Brown, a logger and forester, is from the fifth generation of a family that cut the great white pines of Wisconsin and Minnesota and moved west to log the Douglas fir. His father worked for Weyerhaeuser. Instead of carrying on the tradition of moving west toward the frontier (there was no frontier left) Brown moved east to Maine. "I'm using some of the wealth generated from those forests to do a little restitution," says Brown. He is managing around 300 acres of his own land, called "Steadfast Farm," in Cambridge.

Sam's emphasis is on what he calls "low-impact forestry," and he has worked hard to develop the equipment and methods to achieve his goals. Three key elements in his logging system are his articulated-framed, tracked Dion forwarder, his radio-controlled winch, and his road system. Sam can minimize logging roads and skid trails by cutting narrow paths for his forwarder every 150 feet. This compares quite favorably with mechanical harvesters that need trails every 40 or so feet. Heavy equipment can damage tree roots through rutting and compaction, so the fewer the trails, the better.

Sam practices a short-wood yarding system. Instead of putting out trees whole (branches, tops, and all) as do mechanical harvesters and grapple skidders, Sam limbs the trees where they fall, using efficient techniques learned from Scandinavian expert Soren Erickson. He then uses a radio-controlled cable winch to drag the logs to his forwarder, which is equipped with a loader. The logs are bucked to the most practical lengths for the best markets before loading.

The trailer, onto which the logs are loaded, is also tracked and it is powered by his crawler's transmission. This means that he is not dragging the load, but carrying it. The vehicle's tracks do less damage to the ground than a wheeled tractor in general--but they can still be destructive to root systems of trees.

The machine is only six feet wide, but Sam clears brush and branches a few extra feet on either side to avoid tangling. These trails, which he cuts on the contour, and from which he loads his trailer, are unobtrusive. With his short-wood system, he reduces the wasteful and destructive yarding areas associated with whole-tree logging. Whole-tree yarding areas, in contrast, are at least one, and sometimes two, tree lengths in width, plus the width of the road.

Sam has put around $30,000 into his machine (which he bought used). If he were building it now, it would cost more. Such a cost means it would be prohibitive for a small operation of only a few hundred acres, even though it is quite appropriate for small woodlots. Sam does cut on other people's lands. Sam's system is intermediate between a small farm tractor or horse and the larger equipment commonly used in the industrial forest.

The expense of his equipment compares favorably with a skidder, and is a fraction of the cost of whole-tree harvesters and grapple skidders or the Scandinavian mechanical harvester/processor/forwarder systems. Of course, he also cuts less wood with his system (4 or 5 cords a day, working alone) than these other systems can, but he would rather see more people working with smaller, less-destructive machines, than just a few people working on the larger, more expensive machines.

Sam is amazed at how easily banks will give out loans for the more expensive, more destructive equipment--as long as the borrowers can promise the necessary cash flow. "To pay

[1] First Published in the *Northern Forest Forum*, Vol. 3, No. 1, 1994

off those loans you have to work those machines 26 hours a day," he suggests. "The short-term loans don't encourage long-term management."

Sam uses a variety of winching aids to minimize damage. Instead of wrapping a chain around the log to be winched, for example, he uses a grapple. With the grapple he does not have to lift or roll the log, and, unlike a chain, the pulling force during winching is at the center of the log, rather than at the top. He occasionally uses self-releasing snatch blocks, attached to trees, when his trees are at a difficult angle or when he has to winch around objects.

Sam sometimes uses a smaller machine, called a "radio horse," for prebunching logs for the forwarder. The radio horse also has a a radio-controlled winch, which can be used with multiple snatch blocks to concentrate thinnings from a wide area without having to move the machine. To get the radio horse where he wants he, he pulls out the cable, hooks it to a tree, and then winches the machine along on its runners to the tree.

At the time of this profile, in 1994, Sam was cutting trees marked by his forester Joachim Maier, who is originally from Germany. Sam now has a forester license and does his own marking. Joachim's immediate management goals for stands like Sam's are to remove poor-quality, high-risk trees and to give residual trees optimal spacing for growth and health. He noted that in Germany, foresters in the past often made their forests too perfect. "Every tree was a board painted green at the top," he said. The forests there lacked "crummy trees." "A healthy forest needs some sick trees," observed Joachim. Dead trees or rotten trees are important habitat for numerous insects, birds, and fungi that are essential to the maintenance of forest health. Ironically, "sanitized" forests, where every tree looks perfect, are unstable.

In Maine, Joachim observed, there is no shortage of poor-quality trees suitable for wildlife. There is, however, a shortage of high-quality trees that have the potential to grow big and old. The forest has been highgraded (taking the best and leaving the rest) repeatedly. He and Sam are trying to reverse this process.

Both Joachim and Sam have visited old-growth sites in Maine and want their managed forest to have similar characteristics in terms of height, size, and volume. With their light removals of poor-quality trees and their spacing of good-quality trees, they expect to enhance growth toward more ecologically-desirable stand structures. Joachim hopes that with such treatment, he can have 100-year-old trees attaining the size one would associate with a 120-year-old tree in an old-growth stand. While he is striving to minimize damaging impacts to the soil and to residual trees, he is striving to maximize the impact of stand improvement.

Although aesthetics is not the primary goal of managing for diversity and quality, it is certainly a by-product. In the winter, cross-country skiers use Sam's skid trails and admire the beauty of the forest.

Despite myths propounded by some industrial foresters, doing light selection cutting has not resulted in major blowdowns. It also has not bankrupted Sam, though, admittedly, the initial cuts of small volumes of low-quality wood are not making him rich either. Those initial cuts and his road system were ecological and economic "investments." Wood prices, however, are rising. "I'm now breaking even. I'm paying for my costs and my taxes." After 20 years of cutting, the percentage of high-quality wood per acre has markedly increased over what it was when he started. Subsequent cuts will be of high-quality wood with high-paying markets. "The long-term returns look good," says Sam.

Low-impact logging in 1919...

One horse yarding road

Two horse yarding road

Original captions for photos (From Forrest Colby, *Forest Protection and Conservation in Maine, 1919*):
"Where one horse only is used in yarding timber the saving of small growth and undersized trees is very apparent. Only a narrow path shows where the logs have been hauled to the yard and there is but slight disturbance of the unmerchantable stuff, which is thus left to grow and mature."
"Compared with the one horse road this [the two horse road] is a regular highway from eight to twelve feet wide. Where timber stands thick and there is considerable swamping to be done a large part of the territory actually yarded over is entirely laid to waste."

With modern logging, feller-buncher trails can be fifteen or sixteen feet wide. Eight- or ten-foot wide trails are, ironically, now considered "low-impact"!

5. Residual Stand Damage

Stand Damage Assessment

Stand damage assessments are important for both landowners who want to assure that the logger is doing minimum damage to residual trees and soil and for loggers who want to learn how well they are doing. Without a systematized approach, landowners and loggers certainly can sense that an operation is acceptable or not acceptable, but can not quantify their gut feelings. Being able to quantify stand damage is important if this criterion is to be used in a logging contract. If there are going to be incentives or disincentives for minimizing stand damage, stand damage assessments are crucial.

The Black House stand

On a wet, October 8th (1998) day at the Black House in Ellsworth, the Low-Impact Forestry Project hosted University of Maine professor William Ostrofsky to demonstrate a technique for assessing stand damage. The stand to be assessed had been cut by Sam Brown during a LIFP-sponsored workshop in May. Spring and early summer are particularly difficult times to avoid stand damage, due to the looseness of tree bark, so we were quite interested to see how Sam did. Sam had cut the trees with a chainsaw, using directional felling. He had limbed them in the woods, and winched them, using his radio-controlled winch, to 10-foot wide trails that were spaced 150 feet apart. Most of the wood was forwarded to the yard by Sam's Dion forwarder, though some was forwarded by a Kubota tractor with a trailer and loader.

Forester Teresa Davis's plan called for leaving a well-stocked stand with much pine and spruce. The narrow, dispersed trails helped to minimize the area of ground disturbance. We did not do a soil-impact assessment (but see chapter 7 for an example of how to do this).

The method

To determine stand damage with Ostrofsky's method,[1] we used a five-foot stick to sample all the trees in a ten-foot swath down lines that strive to be perpendicular to most of the trails. We were not perfectly scientific in our approach. We should have mapped the site to determine where to put our lines. Instead, we chose lines more at random. Our prime interest was to learn the assessment method.

Every tree in our ten-foot swath was recorded for species, diameter, and degree of damage.

Damage ranged from:

- "light" (where there were small wounds to trunks or roots on one side of the tree, impacting less than 1/8 of the tree diameter and not going completely through the bark to the wood below), to
- "moderate" (where there is visible damage to tree roots, scrapes on two sides of the tree, or damage of up to 1/4 of the stem diameter), to
- "severe" damage to more than 1/4 of the stem diameter or damage to more than 1/4 of tree roots.

Light damage poses little risk of mortality or serious product degradation. Moderate damage does not pose a serious risk of tree mortality, but will, likely, lower product value and could

---

[1] See appendix II for a sample residual damage measurement worksheet

lower productivity. Severe damage could cause mortality in three to five years and will certainly degrade future product value.

Ostrofsky pointed out to us that tree root damage can be more serious than stem damage at lowering tree growth productivity. He also told us that different tree species and stand conditions have different degrees of vulnerability to damage. White birch, for example, is very vulnerable compared to most other species. Root damage is much less likely in winter when the ground is frozen.

Results

A minimum sample should have at least 100 trees. We miscounted by a few trees, but came in close to that number. Our results? Around 3% of the trees had severe damage. All of this damage was done by the forwarder to trees along the narrow trail. Either the stakes hit a tree trunk or the tracks scraped bark off tree roots. Two percent of the trees had moderate damage--again from loss of bark on tree roots along the trail. Seven percent of the trees had light damage. In contrast, researchers have reported that a "common" skidder operation might damage 20% of the basal area.[1]

Conclusion

Simply because trees are damaged does not mean they must be cut now. If they are expected to live until the next cut and they serve important ecological functions--providing shade or structure to the forest, for example--they might be left. That decision is up to the forester and logger.

The damage assessment exercise with Professor Ostrofsky was eye opening for me and for the others who came along. Simply using low-impact equipment does not guarantee a perfect cut. Both Ostrofsky and Andy Egan (also a professor at the University) stressed the importance of the attitude of the operator, not just the design of the equipment, in avoiding stand damage. Ostrofsky would like to see all operators try to improve their performance, even ones who are doing little damage now. Stand assessments provide operators with a benchmark to which they can compare their future harvests to see if there is improvement. Lowering stand damage is the key to making long-term forestry more viable.

*****

---

[1] *A survey of logging damage to residual timber stands harvested for wood biomass in southern Maine*, by W.D. Ostrofsky and J.A. Dirkman, CFRU Information Report 27, CFRU, UM, Orono, 1991.

Practices to Minimize Residual Tree Logging Injuries[1]

Harvesting is the necessary and critical part of forestry by which numerous products required and demanded by society are obtained. As long as forests are managed for the extraction of wood products, the potential exists for trees and stands to be mechanically damaged. Unlike pathogens, insects, or undesirable climatic conditions, injury to residual trees is one forest health factor under substantial control by landowners, foresters, and harvest operators.

All harvesting does not result in unacceptable damage levels, nor does it often result in widespread stand decline. However, constant attention must be paid to avoid those factors or conditions which can result in unacceptable levels of damage to forests by harvesting activities. Consider the following:

1. Plan skid trails and layout landings.
2. Know the site and stand characteristics (see chart).
3. Assess current (pre-harvest) stand health and tree vigor.
4. Clearly identify the crop (residual) trees--use extra precaution when working near them.
5. Use branches/slash in trails as a protective roadbed.
6. Use bumper trees--designate them before harvesting begins; consider "artificial" bumpers.
7. Consider season of harvest--usually less damage during winter (frozen) months.
8. Choose equipment to match site and stand conditions.
9. Multiple injuries are especially damaging--know pattern of previous harvest.
10. Bark is easily injured during spring and summer; "tight" bark develops quite late in summer.
11. Use high-flotation tires/tracks on the more fragile sites, as appropriate.
12. Limit or concentrate machine activity on skid trails and access corridors.
13. Increase awareness of consequences of mechanical injuries to trees and forest stands.
14. Use silvicultural prescriptions which concentrate harvesting activity, as appropriate.
15. Landowner, forester, and harvesting contractor share job performance responsibilities.
16. Agree to be more "weather sensitive" when harvesting--delay early harvesting if too wet.
17. Minimize the number of stand entries.
18. Recognize that sapling/pole stages are most vulnerable to damage (size and time).
19. Prioritize efforts to reduce injuries to roots/soil first, upper bole/crown next, then root crown.
20. Extraction of heavy loads has higher potential for causing damage than for smaller loads.
21. Use crop tree selection methods rather than area-wide thinning techniques.
22. Avoid harvesting large "wolf" trees (open grown trees with large crowns) whenever possible--girdle and leave for wildlife.
23. Mark skid trail locations prior to harvest.
24. Assess risk of sunscald to residual trees and consider trail/access corridor orientation.

---

[1] Presented at *Exploring Opportunities for Low-Impact Forestry* in Hancock County, May 3, 1997

Prepared by William D. Ostrofsky
Cooperative Forestry Research Unit
College of Natural Resources, Forestry, and Agriculture
University of Maine, Orono, Maine 04469-5755
Phone: 207-581-2877; E-mail Ostrofsk@UMENFA.maine.edu

Natural Factors Affecting Frequency and Severity of Injuries to Residual Trees During Harvesting Operations[1]

	Injuries more frequent and/or more severe	------------------------>	Injuries less frequent and/or less severe
SOIL FACTORS			
TEXTURE	CLAY	LOAM	SAND
DRAINAGE	WET	MODERATE	DRY
DEPTH	SHALLOW	MODERATE	DEEP
ROCKINESS	ROCKY	MODERATE	NOT ROCKY
STAND FACTORS			
DENSITY	HIGH	MODERATE	LOW
AGE	OLD (>50)		YOUNG (<25)
STRUCTURE	UNEVEN SIZED	2-3 CLASSES	EVEN SIZED
TREE FACTORS			
SIZE	LARGE (>8”)	MEDIUM	SMALL (<6”)*
ROOTING HABIT	SHALLOW	INTERMEDIATE	DEEP
SPECIES	INTOLERANT**	INTERMEDIATE	TOLERANT
VIGOR	LOW	MODERATE	HIGH

* Survival of small trees is threatened by major injuries, but if the tree survives, loss in quality will be less than from similar injury on large trees.

** An exception is beech

[1] Table adapted from: Ostrofsky, W.D. 1988. “Improving tree quality and forest health by reducing logging injuries.” In: *Maine’s hardwood resource: quantity versus quality, markets-management*. Maine Agric. Expt. Sta., Orono, ME. Misc. Rpt. 327. pp29-35.

6. Low-Impact Forestry Approach to Protecting Water and Soil

I. Introduction

Water quality. The Maine Woods does not consist merely of trees. It is dappled with lakes and ponds and veined with rivers and streams. In the unorganized territories alone there are 3,000 lakes and ponds and over 16,000 miles of rivers and streams. These waters are essential to the region's fish, wildlife, recreation, and beauty. Our water resources have been called Maine's "life blood."

A key to protecting water quality is protecting the soil. The highest quality water is that which is filtered through the duff and topsoil of undisturbed forests. Where large areas of forest have been cleared for development and agriculture, water quality has suffered from siltation, nutrient leaching, chemical pollution, increased temperatures, and increased extremes of water flow.

Although nearly 9/10ths of Maine's land is classified as "forest," this forest can be disturbed during logging practices in ways that can also affect water quality. Logging roads, for example can be a source of siltation. In the unorganized territories, since 1972, the area of land converted from forest by construction of logging roads was nine times greater than the area taken up by new houselots. Skid trails and heavy logging can also impact soil productivity and water quality. Between 1982 and 1995, logging was conducted on around 7.3 million acres. On nearly half of these acres, loggers removed more than 60% of the stand. The degree to which this activity has affected water quality is unknown since neither logging nor water quality have been widely or closely monitored.

BMPs. By federal law, all states with commercial forests must have rules or guidelines designed to minimize non-point source pollution to surface waters. In around half of these states, including Maine, these guidelines are mostly voluntary, although violations of water quality are enforceable. Maine's guidelines, called "Best Management Practices" (BMPs), relate mostly to road construction, woods trails, landings, and stream crossings. They are designed to minimize or eliminate movement of soil from the forest to streams--where the sedimentation can have adverse impacts to fish.

Recent research has shown that when followed, Maine's BMPs can significantly reduce siltation, protecting water quality and aquatic ecosystems. The Land Use Regulation Commission (LURC) has pointed out that besides these ecological benefits, there are economic benefits for those who follow BMPs:

- the road system works well;
- there are lower maintenance costs;
- there is less wear on equipment;
- hauling is more efficient;
- there are fewer work delays;
- the logger has a better reputation; and
- (to the degree that water quality standards are enforced) money is not spent paying fines.

Why this discussion? Although recent studies have found that BMPs can reduce siltation and other water-quality problems, they also have shown that our state BMPs are not well followed.

FORAT (a Forest Advisory Team that reviewed studies done on BMPs) gave grades (A to F), based on compliance, to each BMP. Fifty-three percent of BMPs got a D or below in a 1996 review.

Other reviews of Maine's BMPs have concluded that there are problems beyond compliance. Even when followed, some of the BMPs are not always effective at preventing serious siltation problems. For example, those who followed BMPs for sediment barriers, breaks in road grades, and hard bottom stream fords still had around 40% major sediment movement.

Several reviews of Maine's BMPs have concluded that they are not comprehensive enough. They do not, for example, cover soil damage that does not lead to water pollution. They don't cover nutrient pollution. They also do not cover impacts on extremes of water flow. While they look at roads, trails, yards, and stream crossings, they do not look at actual logging practices.

Low-impact forestry approaches. While the state has determined that the most pressing need is to get a higher rate of compliance for existing BMPs, landowners, loggers, and foresters who are concerned about water quality and soil productivity want more guidance to do a better job. Many of these same individuals have been attracted to the concepts and practices connected with low-impact forestry.

The recommendations in this chapter are not intended to replace existing state BMPs. These are still a baseline that all LIF practitioners should follow. But for those who want to better understand principles and practices that can improve on BMPs, these recommendations may be of use.

This chapter relies primarily on several studies (listed in the bibliography) by soil and water experts Janet Cormier and Steve Kahl. It also makes use of existing standards for landowners in the Pacific Northwest and Vermont. And it uses standards followed by the Baxter State Park Scientific Forest Management Area (SFMA). The recommendations are, therefore, based on experience as well as theory, but they are not the final word. More knowledge and experience will undoubtedly lead to improvements.

II. LIF approach to soil and water

The LIF approach to protecting soil and water is to:

- Understand the factors that influence logging impacts on soil and water.
- Whenever possible, prevent problems.
- Monitor and maintain permanent roads, trails, yards, culverts, and bridges.
- Promptly put temporary roads, trails, yards, culvert, and bridges to bed.
- Remediate existing problems (revegetate bare ground, repair damaged structures and roads, and restore forest and watershed functions).

III. Factors that influence logging impacts on soil and water

The key to protecting soil and water resources is to understand what factors can lead to problems. Understanding these factors can help a logger or forester to look at the forest landscape and anticipate, for example, what might happen after a heavy rain or a rapid snowmelt, or what would be the consequence of using a certain piece of equipment on a specific piece of ground during a particular time of year.

Degree of soil disturbance. Exposed soil is more apt to move with water or wind than soil held together with fine roots and covered with duff. When soil is rutted or compacted, this closes the pores between soil particles, impeding absorption or percolation. Instead, the water runs on the surface, taking soil particles with it.

Frequency of disturbance. If the soil has been exposed or compacted, it takes time (ranging from years to decades) to reach a more stable state through revegetation and recovery of soil structure. If the soil is disturbed frequently, it may not be able to recover stability and will be less productive and more apt to erode.

Degree of canopy opening and biomass removal. Opening the canopy removes shelter from rain and shade from sun. Direct impact of rain on soil can loosen soil particles and hasten erosion. Direct exposure to sunlight can cause the soil to heat up, leading to a more rapid breakdown of organic matter as well as heating of water. To the degree that there is little plant growth on the site to take up the available nutrients, these nutrients can be leached out of the soil, to flow into the watershed as pollutants.

Soil type. The stability of the soil depends on such factors as: texture, drainage, erodibility, and depth to limiting factors. Knowing soil types is important for planning the location of roads, trails, and yards.

Steepness and complexity of slopes. Steep areas are more subject to rapid runoff and erosion. Long slopes present serious challenges for road and trail construction.

Season. Spring breakup and fall freeze-up lead to conditions where rutting and compaction can reach unacceptable levels. In contrast, frozen soils and dry soils are much less sensitive. These simple facts should govern the timing of operations.

Watershed characteristics. The degree to which logging operations influence a watershed depend, not only on the size and intensity of the operation, but also on the watershed itself, such as its size, or runoff, or seepage amounts and patterns. Also important is the location of the logging operation within the watershed. Is it near a headwater stream, river, lake, uplands, lowlands, bogs, etc.? Some areas, such as headwater streams, are more sensitive to smaller disturbances, for example, than large rivers.

IV. LIF approaches to soil & water problems

Equipment. LIF practitioners match machinery to the size of wood to be cut and hauled, the soil conditions, and the silvicultural methods employed. They strive to use equipment with low ground pressure. They give preference to carry, rather than drag, wood to the yard and to use a loader, rather than a dozer blade, to pile wood in the yard. LIF works best with short-wood logging systems.

Openings. LIF practitioners try, where possible, to leave a relatively closed canopy to protect soil from direct rain or sunlight. They leave slash in the woods, and do not remove an overstory unless there is an established understory, thus ensuring soil cover.

Riparian zones. Riparian zones, the eco-zone between land and water, have many important functions:

- They have a high diversity of plant and animal life.
- They contain important late-successional habitat.
- They are nesting sites for many bird species.
- They provide shelter to streams to help keep them cool in summer.
- They act as filter strips to protect stream water quality from logging disturbance.
- They can (if large enough) provide corridors for animal movement (such as migration and dispersal).

LIF practitioners maintain a no-cut strip (25 ft. to more than 100 ft.) as well as a riparian management zone. The management zone, where light cutting can be practiced, will vary in width, depending on the terrain and wildlife use. The Scientific Forest Management Area (SFMA) of Baxter State Park, for example, looks at factors such as:

- break in slope from upland;
- change in forest type from upland;
- evidence of travel pathways;
- intact developed structures providing connective pathways;
- aesthetic sensitivity to recreation;
- uniform forest structures coincident with wetlands or heath bogs;
- obvious concentration areas for wildlife and wetland habitat.

For a riparian zone to be effective habitat for travel corridors or nesting for interior species (species that do not thrive in edges), the zone may have to be 700 to 1000 feet wide if it is surrounded by heavily disturbed forest. LIF practitioners, however, will strive to maintain at least 75% of full crown closure in the riparian management zone, and at least 65% out of this management zone (assuming the overstory is manageable), so negative edge effects should be minimized.

Managers should keep major roads out of riparian zones, including riparian management areas, except where stream crossings are unavoidable. Trails should be at least 75 ft. away from the no-cut zone. Since small streams are most susceptible to damage from siltation, heating, extremes of water flow, and nutrient pollution, extra care should be taken in riparian management zones to minimize soil exposure, rutting, or major forest disturbance.

Roads. Since roads, with exposed dirt surfaces, are a major source of siltation from logging practices, the first LIF guideline is to minimize road construction. Because trails do less damage, LIF practitioners will forward wood longer distances rather than build more roads.

Managers should keep roads and rights-of-way as narrow as practical. This means, for lesser-traveled roads, using single-lane roads with turnouts. Even for double-lane roads, the travel surface plus right-of-way should be under 33 ft., unless problems in terrain make this limit impractical.

Use of temporary winter roads (on frozen ground) is another way to avoid more disturbing road construction. Winter roads work best, however, on fairly level ground with account taken for drainage patterns.

Some of the more severe problems with roads are associated with long slopes. Roads should be placed on stable, higher ground, avoiding slopes greater than 10%. Wet areas should be avoided, and use of roads during unstable periods when rutting is likely (such as spring thaw or fall freeze) should be avoided as well. Long slopes should, if possible, be broken up with dips. Low spots need to have adequate drainage. To avoid culverts, the road builder can ditch out and outslope.

LIF practitioners should prevent ditches from draining directly into streams. To avoid this, the road builder can use settling ponds or check ponds. These should be at least 75 ft. from the stream. These structures need to be maintained (settling ponds fill with silt).

Newly-built ditches should be revegetated as soon as possible. Ditches with a slope of more than 2:1 are hard to stabilize. Hay bales and silt fences are more effective at *preventing* problems than fixing ones already started.

Permanent roads need periodic inspection and maintenance.

It is better to use old roads (that may not meet all the above standards) that have stabilized than to create new roads. If the old roads have not stabilized and they are poorly sited, they should be revegetated and restored back to forest.

Yards. LIF practitioners minimize yard size by using, when possible, a forwarder with a loader and "hot yarding" (having the truck come when a load is ready rather than having many truckloads sitting in the yard). The yards should be smaller than 1500 square feet. Sloping ground, and low, potentially wet areas should be avoided.

LIF practitioners do not do whole-tree logging, thus avoiding piles of logging slash and the need for large areas to accommodate piles of whole trees and all the equipment necessary to process them.

Trails. LIF practitioners try to minimize exposure and compaction of soil by staying on trails, and by minimizing the size and distribution of trails. Trails should be kept under 10 ft. wide (unless use of larger equipment is necessary due to large wood, in which case the trail should be under 12 ft. wide.) Trails should be 100 or more feet apart, so that trails take up 10% or less of the logging area.

Ground logging systems should not be done on slopes greater than 30%, and trails should be avoided on slopes greater than 15%. Where short sections on steep slopes cannot be avoided, it is safer to run the machine straight up, rather than dig into the slope (creating the potential for more erosion) and risk tipping over as well. Skid humps should be used to break up long slopes.

Loggers should plan ahead to locate trails on higher, more stable soils. Old trails that have stabilized are preferable to creating new trails, but old, rutted, eroded trails should be avoided and should be revegetated.

Loggers should avoid using trails when conditions are unstable and there is a risk of creating ruts that extend beyond the *A* horizon (the soil horizon just below the organic pad and where organic matter and minerals from parent rocks mingle). Rutting can sometimes be avoided by placing a mat of slash on the trail--but in wet conditions, even this may not work with heavily loaded machinery.

Stream crossings. Because 1st and 2nd order streams are most sensitive to disturbance, stream crossings should be avoided, when possible. The SFMA uses streams as boundaries for its management units to help avoid a need for crossings.

It is difficult to avoid some siltation when making stream crossings, so extra caution is required. The builder should use the crossing that causes the least disturbance to the stream. SFMA uses wooden bridges for 3rd order streams to avoid disturbing the stream bed.

Roads should approach bridges at a right angle to the stream. The road should slope up, if possible, just before the bridge, or there should be dips to avoid direct discharge from the road into the water. The road + right-of-way should narrow before the bridge to minimize exposed soil surface near the stream. A one lane bridge for less-traveled roads is preferable.

Permanent bridges should be designed to withstand 100-year flood peaks. Wood should be carried, rather than dragged, across bridges to avoid dragging dirt that will silt up the water.

Permanent culverts should not be put in during winter because there can be problems with frost heaving and settling. Use of gravel under the culvert can help to avoid settling. Rip rap and seeding have been used successfully to stabilize culverts.

Restoration. Temporary roads, trails, and crossings must be promptly put to bed. This has been a major weakness in many logging operations. The logger should restore, where possible, natural drainage patterns. Exposed soil, especially on slopes, should be revegetated. On slopes, the logger needs to account for cross drainage, using water bars. Temporary bridges and culverts should be pulled up before high-water season.

V. Conclusion

While these recommendations can help improve the performance of logging operations to protect water and soil resources, experience will lead to further refinements and improvements. Indeed, it is not often that road building or other logging enterprises will go right the first time--refinements and corrections will probably be needed over time. Loggers can benefit not only from their own experience, but from observing the performance of others. Soil scientist Janet Cormier found some very thoughtful innovations done by loggers who understood the principles of water protection and applied them to specific circumstances. One can also learn from mistakes. Certain "good" ideas may not work in all situations.

For further information (including pamphlets with illustrations of BMPs), consult the resources below:

Resources

Bissel, Jensen, 1998. Baxter State Park Scientific Forest Management Area management plan (draft, 60 pp.)

Briggs, Russel, Alan Kimball, and Janet Cormier, 1996. *Assessing Compliance with BMPs on Harvested Sites In Maine: Final Report.* CFRU Research Bulletin 11. University of Maine, Orono.

Cormier, Janet, 1996. *Review and Discussion of Forestry BMPs.* Maine DEP and USEPA.

FORAT, 1996. Background, mission statement and findings on BMPs. MDEP, Augusta. (2 pp.)

Hammond, Herb, 1998. *Silva Forest Foundation standards for ecologically responsible timber management (May draft).* Slocan Park, British Columbia (87 pp. plus appendices).

Kahl, Steve, 1996. *A review of the effects of forest practices on water quality in Maine*. University of Maine, Orono/DEP Augusta.

Land Use Regulation Commission, 1995. *Erosion control on logging jobs.* DOC. Augusta.

Maine Forest Service, 1992. *Best Management Practices Field Handbood.* DOC, Augusta.

Vermont Family forests, 1996. A voluntary timber management checklist. (6 pp.)

Wiest, Richard L. *A Landowner's Guide to Building Forest Access Roads.* USDA Forest Service, Northeastern Area, NA-TP-06-98, Radnor, PA, 1998. 45 pgs.

7. Soil Damage Assessment System

Soil is the foundation of the forest. It is where the roots dig in. Soil not only holds trees up, it also supplies water and nutrients for the trees. Damaging soil can lead to siltation of water, regeneration problems, undesirable shifts in species, and lowered productivity. With low-impact forestry, we try to minimize such damage. Favored low-impact logging systems try to keep distribution and area of roads, yards, and trails to a minimum.

With the death in 1997 of soil scientist Janet Cormier, the Low-Impact Forestry Project lost a priceless resource on soil information. Janet did an important study on Best Management Practices the year before, and only a few months before her death from cancer she participated in a LIF conference in Ellsworth, describing her results.

BMPs are designed mostly to prevent siltation of water. They are not designed to minimize soil damage in general. In 1998 we approached Jim McLaughlin, Assistant Research Professor of Forest Resources, Cooperative Forestry Research Unit, University of Maine, to see if he could come up with a system for assessing soil damage. We had already benefited from the expertise of Jim's colleague at the CFRU, Bill Ostrofsky, who had demonstrated to us a system for assessing damage to residual trees. We asked Jim to come up with a numerical rating system, similar to Ostrofsky's.

Jim came up with a system that he first demonstrated in Vermont in late April 1999, at a conference put on by Barbara Alexander of the Vermont Loggers' Guild. In early June, Jim took five members of the Maine Low-Impact Forestry Project out to the University Forest to demonstrate his system to us.

Assessing Soil Hazard

To accurately measure soil damage, one has to do a pre- as well as a post-logging assessment. The pre-logging assessment is key to preventing damage. With the McLaughlin system, the forester determines potential hazards by examining the climate, soil types, hydrology, and drainage of the area to be logged. The forester lays out transects, digging occasional pits to determine depth to soil restricting layer and soil texture. The forester also determines topography and slope.

Jim has come up with a point rating system for these categories. With a high hazard rating, the logger has to take special care—including restricting logging to certain times of year. Jim recommends the following restrictions, based on drainage class type:

Operational Constraints

Period of optimum operability for skid trails for forests of Maine

Drainage	Period of Optimum Operability
Excessively	January through December
Well	May through February
Moderately Well	June through February
Somewhat Poorly	July through September
Poorly	July through September, frozen soil
Very Poorly	On frozen soil

Post logging assessment

Jim has developed systems to measure the area of land taken up in roads, yards, and trails as well as the degree of damage to soils. As the area in roads, trails, and yards increases, the area of productive forest decreases. Much logging damage occurs near the trails—especially to tree roots. In Scandinavia, landowners expect loggers to have no more than 20% of logging areas in trails. By hauling cable and setting trails up to 150 feet apart, loggers can have less than 10% of the land area in trails.

To determine the percentage and degree of soil disturbance, the forester lays out transects (perpendicular to trails if possible) with periodic survey points. When the forester stops at a "point," he or she then scans an area approximately 36 square feet around that spot. The following chart summarizes the classes of soil disturbance in the McLaughlin system:

Class 1	Class 2	Class 3	Class 4
Undisturbed	**Slightly disturbed**	**Deeply disturbed, surface soil removed and subsoil exposed**	**Rutted, compacted**
	Litter in place Litter removed and mineral soil exposed Mineral soil and litter mixed Mineral soil deposited on top of litter		0 to 6 inches 6 to 12 inches >12 inches

What is needed next is a spread sheet form that will allow foresters to check off criteria as they do their surveys to quickly come up with damage ratings. After that, the next step is to do a number of surveys on a variety of logging sites to determine what the numbers mean. What ratings are unacceptable? What are acceptable? What are desirable? Once this has been worked out, Low-impact foresters can better assess logging operations using McLaughlin's and Ostrofsky's systems. Landowners need to know if loggers are living up to desired standards. Landowners can use these assessments to find out if the logger qualifies for a bonus for high-quality practices. These assessment systems will be crucial in learning what techniques and technologies yield the most desirable results in given situations.

*****

8. The Living Soil[1]

Soil scientists normally have an expertise in the physical characteristics of soil--the type, the horizons, the chemistry. They often do not have as much knowledge about the life *of the soil. In the following excerpt from an interview from 1998 (see introduction to section on Ecosystem Management), forest ecologist David Perry talks about some of the complexities of life in the soil. The below-ground ecosystem can be as complex as or more more complex than the above-ground ecosystem. Compromising the complexity of life below the ground can have serious impacts on what lives above the ground.*

BA: You've described the soil-plant relationship as a "dance of mutual creation." Would you describe the role soils play in maintaining a stable ecosystem and explain how breaking the link between plants and soils can affect potential site productivity?

DP: Plants need about eighteen elements as nutrients; microbes and animals need another eight or nine. All but two of those come from the soil. So the soil is the repository of most of the building blocks for life. The soil is the bank that holds the water between rainfalls and that modulates it out either to streams or to the air in an orderly way.

Soil is, by far, the habitat of the greatest number of organisms in systems, probably 50% of the animal biomass in forests is below ground. The food chains below ground are generally long, more complex, more diverse than the food chains above ground. All of that life below ground in soils, and the physical structure of soils that gives them this capacity to store water and yet to drain water and to let air in, comes from energy that has been put in by plants. Plants pump about 50% of the energy they capture in photosynthesis below ground. Some people believe it's more than 50%, and I'm one of those people.

So a lot of the energy that's gathered by plants goes below ground, and it builds soil structure and feeds organisms, and a good share of these organisms are doing things that feed back in a very important way to the growth of the plant: they're cycling nutrients, they're gathering nutrients and water, they're helping to protect the plant against pathogens. That's what I was referring to when I said that plant and soil become joined together in a dance of mutual creation. The soil ecosystem depends on the energy from the plants and the ability of the plants, to gather that energy depends on a fully functioning soil ecosystem.

Forests everywhere are disturbed to one degree or another. We talk about ancient forests, and they may be ancient, but they're not immortal. Trees die. Sometimes large-scale disturbances come and wipe out large numbers of trees. One of the primary mechanisms of resilience is the ability of plants in a system to recover very quickly; and often they're a different set of plants than the one that we focus on commercially. They're hardwood shrubs that can sprout from roots or that store seed in the soil where it's triggered to germinate after a disturbance. Or they're annual plants whose seeds are dispersed widely on the wind--a whole collection of things that come up and stabilize the soils. And they maintain soil integrity as the system recovers.

One of the classic studies in ecology was done on Hubbard Brook Experimental Forest back in the late 1960s. They logged some areas, herbicided all of the early-successional vegetation, and then looked to see what happened in the streams. What happened was predictable, but it had never been demonstrated quite so elegantly before. When you knock out that early-successional

[1] First Published in *Northern Forest Forum*, Vol. 6, No. 3, 1998

recovery mechanism, nutrients start bombing out of the system, soil integrity is disrupted. A lot of nutrients are lost, and it may take a long, long time to build those back up again.

It's an interesting thing--there has been much talk among ecologists over the last ten years or so about the importance of disturbance in systems and how changeable natural ecosystems are; and that's certainly true if you look above ground. Over decades and centuries there is quite a bit of flux in ecosystems, and in the plants that are present, and so forth. But if you look below ground, things change much less, they tend to be more stable, and those points of stability are what confers the ability of ecosystems to hang onto integrity and reform themselves over time after disturbance.

BA: I wanted to ask you about the role of mycorrhizal fungi.[1] You've already addressed this issue when you were speaking about the exchange of nutrients and the protection from pathogens in the below-ground ecosystem. But how key a role is this?

DP: It's absolutely essential. There are very few plants in the world that can grow successfully without mycorrhizal fungi. All forest trees in the temperate and boreal forests, and probably all of them in the tropical forests as well, require mycorrhizal fungi. The mycorrhizae gather water and nutrients, protect against pathogens, and extend the life of feeder roots. There may be thousands of times more gathering surface in the hyphae of mycorrhizal fungi than in roots. By far, the greatest presence of trees below ground is manifested through the mycorrhizal fungi.

It's now clearly established that mycorrhizal hyphae link trees of the same and different species, and the nutrients and carbon move between trees through these linkages. When you look above ground, you see a bunch of individuals. When you look below ground, that individuality becomes less clear.

The fungi do another thing. We're coming to believe that they confer a great deal of stability on the system because they're so diverse. If you take one tree, it's a single genotype, but that tree may have anywhere from twenty to fifty different species of mycorrhizal fungi on it. So the diversity of that one genotype all of a sudden gets manifested out into a minimum of twenty or fifty different genotypes that are present in its symbionts. When you account for the genetic diversity that is likely to exist within a given species of fungus on a tree, that diversity is magnified even more.

The fungus has a great deal of evolutionary capability over a short period of time that trees cannot have. Trees, being long-lived organisms, may produce a seed crop every two or three years, but how frequently do the conditions exist so that they can have progeny that will succeed? That may be a period of years or decades, and, in some forests, like the boreal, over a hundred years. Being long-lived organisms, trees have a fairly slow response capability in an evolutionary sense, and that puts them at a great disadvantage in their evolutionary sparring match with tree-eating insects and pathogens, which can turn generations around rapidly, and therefore evolve very quickly.

Yet, if you factor in the tree's symbionts--the mycorrhizal fungi, the foliar endophytes[2] [...]--then an evolutionary capability is conferred on the tree, through it symbionts, that puts it on

---

[1] Fungi that form extensions of plant roots and increase nutrient and water intake.

[2] Foliar endophytes are microfungi that live symbiotically in plant leaves and help protect their hosts against pests and pathogens.

more of an even par with the pathogens and the tree-eating insects. That is a little explored, but potentially very important, aspect of the tree's symbiosis with fungi.

It's an interesting thing, if you look at both of the major surfaces with which trees interact with the environment--the crown and the roots--both are characterized by symbioses with fungi.

*****

9. Logging Labor Issues

Understanding the history of how logging technology and labor practices developed can give us perspective on what we now perceive as "normal." These technologies and policies were developed for industrial-scale logging--to have high productivity in removals for the lowest cost. Current approaches to loggers are not especially conducive to low-impact forestry nor have they been beneficial to local communities. While these policies may be the status quo now, they are surely not set in stone. They can be changed to better fit a forestry model that cares more for forests and communities.

For centuries, logging techniques in Maine were fairly stable. The trees were cut with axes and saws, taken to a yard with horses or oxen, then floated down streams and rivers to waiting mills. It was seasonal work--most of the cutting was done in the winter--the farmers' off season. This basic system persisted even as sawlogs were replaced with pulpwood. The big change in technology came after WWII, with chainsaws, skidders, logging trucks (with loaders), and then, eventually, mechanized harvesters. Forest harvesting became "industrialized."

For a while, some of the companies had their own woods camps and hired loggers as employees. Over the last few decades, all the companies have gotten rid of such employees and now hire "independent" contractors--who have to take care of insurance, workers' compensation and other payments themselves. A Department of Labor report in 1999 said that self-employed "independent" contractors "...from the standpoint of US labor law,... do not exist." It also mentioned that loggers and logging contractors are telling their sons to not get into the logging business. "The intergenerational chain which has produced loggers in the woods for perhaps hundreds of years may be strained to the point of breaking.."

Because forest ownership in northern Maine has been so concentrated, companies can pay wages lower than free-market levels--workers can take it or leave it. If they leave it, there are loggers from Quebec who will take it. This has been a contentious issue for Maine loggers to the point that some have blockaded the border.

The issue of imported woods labor is not new to Maine. In 1883, a letter writer to a Portland newspaper complained: "There is much comment on the species of protection that allows hundreds of horses to be brought here openly from New Brunswick, bonded at a small cost, and allowed to work through the lumbering season and return home in the spring...Our lumber is of no benefit to us; it is cut and hauled and driven down the streams and goes back home again. If the province teams were not allowed to come over here and work, our farmers could get employment for their teams.."

During WWII, labor shortages in the Maine woods were so great that companies imported thousands of Canadian workers from Quebec and even used German war prisoners to cut their wood. The use of Canadian labor to cut Maine wood was codified into US law by the Immigration and Naturalization Act of 1952. The Act permits non-immigrant aliens to be employed in the harvesting of agricultural products as bonded labor. "Bonds," unlike labor on visas, are recruited for specific jobs--supposedly if no domestic workers can be found.

Under the Act, there are supposed to be protections to American workers. Employers can only hire if no domestic workers can be found. There should be no adverse effects to employment, wages, or working conditions. In 1975 Maine loggers went on strike. Part of their complaints were about Canadian labor. Despite government promises to fix the problem, real wages (inflation adjusted) have actually declined since then. Fewer Maine workers will work for such wages. To make a "living," workers have to put in 55, 60, or more hours a week. The DOL report acknowledged that the "prevailing wage" the government was setting was well below a free-market wage, but concluded that hiring Canadians had no adverse effect. The report stated that concentration of land ownership was a more significant factor in explaining wage levels. Maine loggers disputed the former conclusion.

The combination of mechanization and Canadian labor has greatly reduced the number of Maine loggers. To some, logging is so dangerous, that reducing the number of those employed is seen as a plus. But, for some reason, employers are having trouble recruiting young loggers into the profession, even as the jobs have become less dangerous. And Canadian labor is still an issue despite promises from paper company representatives during the 1970s that with increased mechanization, the more desirable jobs would be filled by eager Americans.

I interviewed a local logger, Jimmy Potter, in 1997,[1] to explore the changes in logging technology, logging techniques, and payment policies over the years he has worked--which spans the period from the end of WWII to today.

Jimmy lives in Drew Plantation, a town 35 miles north of Lincoln, Maine. He lives with his wife, Marlene, on a hill overlooking Mount Katahdin to the west. He has cut for companies that no longer exist or that have gotten rid of their land in Maine: Penobscot Development Company, Diamond International, Saint Regis, Great Northern Nekoosa, and Georgia-Pacific.

In 1975, Jimmy was a vice president of the Maine Woodsmen Association (MWA), the group of loggers and truck drivers who went on strike against the paper industry. The MWA actually succeeded in shutting down several mills. Industry claimed that these strikers, many of whom were "independent" contractors, were violating anti-trust laws. After a Maine court issued an injunction against the strikers to end "economic warfare," the strike lost momentum, and life in the woods returned to "normal."

ML: How long ago did you start working in the woods?

JP: Well I've been in it roughly 50 years. I started working for money when I was 12 years old.

ML: What were you doing then?

JP: I was limbing with an ax and we was peeling in the summer time.

ML: You were peeling...?

JP: Spruce and fir and hemlock primarily. Back then pulpwood had to be peeled before it got to the mill. This changed sometime in the nineteen fifties.

ML: And you were cutting it to 4-foot lengths?

JP: We peeled and cut all summer and let it lay. When the bark stuck on it in the fall and we couldn't peel it no more we started yarding it with a horse and sawed it up to 4-foot logs, piling it all in a line by hand.

---

[1] First published in *Northern Forest Forum*, Vol. 5, No. 5, 1997

ML: Did you cut the trees down too?

JP: Oh yeah, I was helping to cut with a bucksaw, a crosscut or whatever.

ML: Did the woods look different then?

JP: A lot different. You had an overstory about everywhere you cut. You never had an opening where you'd look right through and see ground under it. You'd selective cut and you left stuff eight or ten inches...you took the biggest trees out. There was a lot of big wood around in those days. We built roads into areas that had little or no previous cutting. I cut spruce that had sixteen, twenty, or even thirty feet before the first limb. We could get over a cord of wood from one tree.

Peeling hemlock. From *Sixth Report of the Forest Commissioner of the State of Maine 1906.*

ML: You were cutting these trees for lumber?

Four foot pulpwood piles. From Ferguson and Longwood, *The Timber Resources of Maine,* 1960.

JP: No, we cut most of it four foot for pulpwood. That's what the markets were for. Lumber didn't come in big 'till they started up the stud mills.

ML: When you first started working in the woods, was woods working a seasonal job?

JP: Yeah, primarily, as far as cutting it. But it was a year-round job, because you'd go out...A good man could go out and probably cut three hundred cords of wood a year and survive on it. You'd peel all summer and would yard the wood in the fall and take your team of horses and haul it to a brook or river and then in the springtime you'd go on a drive and drive it to the mill. So you were working year round on a limited amount of wood.

ML: How was the community different then?

JP: Everybody was working then. They had a job. The farmers had a farm to work on. They'd go cut pulp in the afternoon or cut pulp in the morning. Everybody was working and had money.

ML: Money bought more then?

JP: A lot more. Sixty dollars a week was a big pay week in those days.

ML: Were you concerned with wood supply then?

JP: When we were cutting with a horse, a bucksaw, and an ax, you wouldn't figure you would ever cut all the wood that was growing out there. It looked like there was an endless wood supply. You'd go to work and you'd be in an area for 15 years and you'd almost be able to start all over again where you went the first year--you could cut right through again.

One horsepower forwarder. From Ferguson and Longwood, *The Timber Resources of Maine,* 1960.

ML: What changes started happening as you were getting older?

JP: First thing that came in was the crawler-tractor, and then the chainsaw came along.

ML: What did you think of that when you first saw it?

JP: It seemed like a good thing, you know. It was going to save a lot of back-breaking work. You wasn't going to be pulling a saw and crosscut by hand. You could do a little more. But the thing of it was, the price didn't change that much.

ML: The price of the wood?

JP: The price of wood. Everything stayed the same. Back then I paid five hundred, almost six hundred dollars for a chainsaw in 1953. Five horsepower, thirty pounds. You cut down with that, but you couldn't limb with it. It was a big chain that turned too slow. You still limbed with an ax and you'd saw it up four foot with the chainsaw. That made a big difference on sawing wood. A good man with a bucksaw, if he could saw five cords a day with a bucksaw he had a pretty good day.

ML: Five cords!

JP: That was for a two-man crew just yarding and bucking trees. You take a chainsaw, and a three-man crew could cut ten cords. That includes felling and limbing with no peeling. It increased production.

ML: When did skidders come in?

JP: Nineteen sixty seven or sixty eight.

ML: What did you think of the skidders then?

JP: That was going to be another thing. The skidder didn't get tired. You could do a little more work, work a little longer. Work in the heat. A horse had to have his wind like you did. That's when things started going backwards. A man had to compete with a machine. That's where your injuries came in.

ML: There weren't a lot of injuries before?

JP: No, we never had that many injuries. Most of the while you might get nicked with an ax or something like that. You never had that many bad backs even though it was rough work. I never had nobody have any problems. Everybody took their time and they lifted the way they were supposed to and two men would lift on a stick of wood.

ML: How did mechanization change the economics of cutting wood?

JP: With mechanization, your cost of operation went up. But the price of your product didn't increase. So you had to do more to keep even.

ML: Couldn't you cut a lot more?

JP: Yes, you probably doubled or tripled your output, but you didn't have much more money when you were done than when you were cutting with the ax.

ML: What led workers to organize to form the MWA?

JP: Everybody had different reasons. For myself, I could see that the forest was declining. They would take an area, put in a road and pretty well wipe it out. For some it was on account of the pay scale, for others it was Canadian labor. There were several different reasons.

ML: But a lot of people decided at the same time something had to be done. Were people upset earlier than that?

JP: There was dissension in the woods when they started going with the foreign labor coming in from Canada and taking jobs away that we could do here. And when you got the machinery, you had to work year round and we weren't getting year-round work. We'd buy a skidder and you had a payment for twelve months but you was working eight months. And then there'd be a shutdown, or there would be a quota system put in. If the supply in the summer time got too much in the mill they'd shut back, cut your contract back. Your skidder still costs so much per month.

One thing always amazed me all through the years I worked in the woods--that when the mill price for the wood went up, the guy cutting the wood, back in hand-cutting days, he got the least raise of anybody. It seemed like he was the last to get the increase. The minute the wood dropped down a dollar or a half a dollar or whatever, he got the first cut.

ML: I have been told by contractors that they're not making any more money now per cord than they were in the mid eighties. They say that as the mill-delivered price goes up, their price per cord does not go up--the companies just raise their stumpage.

JP: The stumpage raises automatically. That's been my kick on Tree Growth.[1] If the right numbers were put in the equation, it would be quite a different land value. All the state uses is what I as a contractor cut on my ground or cut on someone else's ground. I gotta report it, but wood cut on the paper companies' land isn't part of the equation that the state uses to fix values. But they are paying taxes as though their stumpage was as low as everyone else's. Even though their average may be a multiple of the state average.

ML: The complaints workers had when they started the strike in 1975...how many of those have been addressed since then?

JP: We're still addressing, I think, the same problems we had back then. Everything that has taken place in so-called forest management has been addressed for a period of probably twenty five years and hasn't changed.

ML: When you say "addressed," you mean people have been talking about it?

JP: Talking about it. And they say we'll do this or we'll work that. When we started out with the MWA, we had an Independent governor, Governor Longley. Since then we've had a Democratic governor, a Republican, and now we're back to an Independent again, and I don't see that there's been a damn bit of change in over all these administrations.

They came in with different things to lead you to believe that there were going to be some changes made. They came in with Tree Growth. We'll give them a break on their taxes, therefore they'll cut less wood off the ground and a sustainable forest will still be here...and it hasn't ever happened.

We had the wool pulled over our eyes with Tree Growth. We didn't kick about having to pay more taxes in our small towns because we were giving the company a break and we helped them out thinking it would help us all out. We'd be happy. We'd be working. We wouldn't be traveling one hundred miles one way to get a job, and we'd work right here at home. It sounded like a reasonable approach. I didn't actually like the idea, but the state was going to reimburse us for some loss...which they didn't do.

So that was the first thing that came in. Propaganda I guess. Every bill they put through...they'll negotiate around until they get a certain amount of wood cut offen a piece of ground--the way they want to cut it. They basically run the state as far as the laws being made.

ML: Since the MWA strike, there's been a big reduction in labor due to mechanization. What has the mechanization done to the woods? Has it improved woods work? Have things gotten worse or are they the same?

---

[1] "Tree Growth" refers to the Tree Growth Tax Law that is a current use tax on timberland that is based on the growth rate and averaged stumpage values (by county) of softwood, mixedwood, and hardwood timberlands times a discount rate.

JP: It's probably improved work for some people, but like you said, it's cut the labor force back. They don't need so many people to get the same amount of wood. When we first had the small skidders, they weren't that bad. They'd take a cord or so at a twitch,[1] and it would be limbed wood. Now they're coming in, getting two cords or two cords and a half, limbs and all, and taking a thirty foot swath out of the forest.

ML: Thirty foot swath?

JP: Just about, when you get all done with it. What you cut is fifteen or eighteen feet wide. And when you grab these hardwoods by the butt and bring them down through the woods, they just fan out. You've got this bare trail where you've already cut with the shear, and then you've got the stuff on the side you're knocking over.

ML: They mean to have a fifteen foot trail, but they kill so many trees along it, they effectively have wiped out thirty feet?

Contractor's choice. Feller buncher trails dominate the stand.
Aerial photo by Mitch Lansky

JP: Most trees will still grow with the bark knocked off of it, but they're going to be diseased trees--they're not going to be a good piece of timber.

ML: Have you seen some landowners who are doing a better job?

JP: I haven't been over a lot of it, but some of it I've seen where they've made an attempt, I guess I would put it, to put a main trail in and wishbone offen it and cut. But to me it still doesn't make a good forest practice to take out such a wide strip of ground for the skidder trail like they do. You've got an open canopy. You don't have a canopy over these trails.

When we yarded out with a horse, you'd have a six foot, five foot trail. The canopy could close above it. It didn't create a clearcut area because it would grow right back. But these wide trails for the machines to me are a clearcut because there's nothing left in them. They cut everything down that can be yarded and they run over everything that's not merchantable.

ML: What are some of the things you see happening because of these open canopies?

JP: You get blowdowns, it's one of the big things.

---

[1] A "twitch" is a pile of logs hauled out to the yard.

ML: Have you seen a lot of blowdowns?

JP: A lot of blowdowns. They don't differentiate what type of forest they're cutting. If they're going in with a shear[1] and cut a piece of ground, they have the same cutting practice if it's ledgy with six inches of soil or if it's two feet deep with loam. They don't make any allowance for what type of ground they're cutting on.

ML: You don't sound impressed with the management.

JP: They take a certain amount per acre off a piece of ground with a diameter cut and leave so much standing. Instead of leaving a good tree that's going to make real good lumber in another ten or fifteen years, they'll leave the stuff that won't make good lumber because they want to make money off the wood they're cutting right now. There's not much in the way of forester supervision, and some of these shear operators are only eighteen or twenty years old and don't know a spruce from a fir.

They often damage the trees in both cutting and limbing so they are lower grade. When they cut to length in the yard, they don't consider each tree for the highest market. They just start at the butt and cut the tree out to sixteen feet, which might go for a few hundred dollars for a thousand board feet, where if they cut the best sections to nine-foot six, it might go to veneer and be worth, say, eight hundred dollars a thousand.

You sometimes have brush piles in the yard twenty feet tall. If you take these limbs and tops back to your skidder trails and tread 'em down, they're three, four, even five feet deep. Cut areas starting out like this will take forever to be regenerated.

ML: They might just use those as trails again. But what they've done is reduce the effective size of the forest, perhaps by twenty percent or more. What does this type of cutting do to wildlife?

JP: We've got plenty of summertime feed for deer here, and we've got moose to live here year round, but the deer population can't overwinter in these sparsely forested areas. They need overstory.

I've also seen impacts on fish. In the old days, if you got a two-day rain, you'd have to wait three or four days for the water to drop to go fishing. Now if you get a couple of days rain, you'd better go out the next day, because two days from then the brook is going to be dry again. There's nothing holding the water in the woods. The skidder trails on the ridges cut off the roots and leave ruts. Water doesn't work its way slowly down like it used to and it's got more silt. The brooks just bounce up and down. It has to be due to the way the forest was cut, because that's the only change you've got out there--that and your road systems.

ML: What would you like to see in the woods if you could control state policy?

JP: My first idea to get back the forest here is to do away, basically, with all the mechanical harvesters--the shears, the delimbers, and the grapple skidders--yarding whole trees. They have to go. There's no place in the woods for 'em if they're going to keep the forest. This would put more people to work in the woods.

---

[1] A mechanical harvester

ML: Would you pay them by the cord?

JP: If they could pay a reasonable price. It would have to work out to a good wage no matter how they paid.

ML: Have you seen any equipment that you think does minimum damage?

JP: I saw what they call a Valmet[1] working on Baskahegan ground. I never watched it work too much, however. They go in and this machine cuts down a tree, limbs it and cuts it into certain sized blocks. They've got another machine called a forwarder that comes along and they take it out. So you're not yarding that tree through the woods limbs and all. This machine is only good for working with softwoods. One of the best jobs I've seen is where some guys yarded to a trail with horses, and then a skidder twitched them to the yard.

ML: Having worked in the woods for 50 years, you're now retired. Would you say that you are making enough through social security to live out the rest of your retirement happily?

JP: I'll go back to one of your questions before. You talked about forming the MWA. That was another thing that we tried to talk about and negotiate with--getting a wage similar to what the people in the mill were getting. We were cutting the wood, supplying the mill, and getting paid by piece rate, while your mill worker was getting paid by the hour. When I first started cutting wood, I was making as much in the woods as the mill worker was making in the mill. Now I'm not making a third of what some of these guys are making. When they retire from the mill after so many years, they have a company retirement plan, plus they can draw social security.

ML: They are employees and you're not.

JP: They are employees. In the very beginning, with my first contracting, I was a self-employed contractor. I bought the stumpage from the companies and sold it to where I could find the best market on some species. The majority of the wood went to the company. They told me what to cut and where to take it. As long as I was doing some of this on my own--buying wood off of the company and selling it where I wanted to--I could say yes, maybe I'm a self-employed contractor. The latter part of it, they told me what to cut, where to cut, and what I was going to get for it. I couldn't negotiate a price.

ML: That sounds like you were their employee.

JP: I was basically an employee.

ML: But you got no benefits...

JP: I got nothin'.

---

[1] A type of single-grip harvester/processor

ML: You got no paid vacations...

JP: No.

ML: You got no good retirement...

JP: No. I didn't make the company parties or cookouts or things like that. I'm just a person I guess, just a number.

When we talked about getting a rate something similar to what they got at the mill, they argued that the mill was classified as "skilled labor," and we were classified as, I guess, "unskilled labor." My argument back then, and still is today, is that I could go in and learn to run a machine or forklift in the mill or pull wood in the mill and do the job. I doubt if very few of them mill workers could go out and take my chainsaw and make a living cutting wood.

ML: What about this practice of paying the loggers so much for their work, so much for their skidder, so much for their chainsaw? Do you think some of your employers abused that?

JP: That definitely was abused. When the workman's comp came in, it was based on your wages. The more they could get allowed for skidder allowance, or chainsaw allowance or travel allowance...that made the actual wages less money. So they had to pay less comp.

ML: But that also means less unemployment if you're out of a job.

JP: And you got less social security.

ML: How much are you getting per month in social security?

JP: Five hundred and thirty five dollars.

ML: Can you live on that?

JP: I can survive on it, but it doesn't support my vices. I've applied to a couple of jobs cutting wood[1] and I can't even make an application out until I take this course to be a Certified Logging Professional, which takes a week and costs about five hundred dollars.

ML: Who pays that cost?

JP: I do.

ML: And you're going to be certified as a professional logger after you've worked in the woods for fifty years after paying a month's worth of your social security? I've talked to people who have worked in the woods quite a few years and they say they have learned something from the course.

---

[1] Not long after this interview in 1997, Jimmy went back to work in the woods. He is still working there in 2002.

JP: Oh yeah. There's always room to learn. I'm not against going taking a course to learn something, because things change. The things you're using change, and you always learn something no matter what you do. To have to pay five hundred dollars to learn to do a job that I've been doing for fifty years doesn't seem to make a lot of sense to me.

ML: Do see any hope for the woods?

JP: I was told one time, I think probably about twenty years ago now, that they were cutting about two and a half percent of the state [per year]. Two and a half percent of the state means it would take forty years to go through the whole forest. Now I hear they're cutting at a three percent rate, which means a cutting cycle of around thirty years. They're cutting some now that we cut fifteen years ago and done a pretty good job. Now they're coming back with shears and just about clearcutting it.

To increase or sustain the cut they're going to have to cut an even wider area. But the way they're reducing the forest with these machines I just can't see how this can be sustained. It's like the fishing industry. It was a big ocean and a lot of fish, but they aren't there no more. And that's what's happening with the trees. Eventually it will be the same scenario as what's going on with the fishing.

*****

10. Paying Loggers

How do you pay loggers to cut wood? The question might seem strange, given that loggers have been paid to cut wood for generations. Why ask now? Because the more common payment systems have neither served the loggers nor the forests well.

Traditionally in Maine, loggers operating chainsaws and skidders have been paid on a piece rate--by the cord. The goal has been to encourage production. Production means getting the wood out fast. The more a logger cuts, the more he makes. Logging equipment has also been designed for production. Today's dominant technologies--feller-bunchers, grapple skidders, and delimbers--can move a lot of wood fast. While the contractors who own the equipment are paid by the cord, some of the operators of these machines are paid hourly wages, like factory workers, although they are paid a fraction of what paper-mill workers get.

But look at the woods. Something has been forgotten in the name of productivity of extraction; the productivity of the residual forest. Too much land gets taken up in trails and yards to accommodate the machinery. Too many residual trees get damaged. Too often the result doesn't look much like a forest. As John Arbuckle said, "You get what you pay for." If you want to reduce residual damage and manage for a well-stocked, high-quality forest, you have to pay the logger appropriately.

The question of logger payment systems is important to the Maine Low-Impact Forestry Project. We are interested in results on the ground that lead to long-term benefits for the landowner, but also compensate the loggers for the efforts to reach such a result. A landowner association could, potentially, improve marketing of wood by using a concentration yard, or even set up value-added opportunities, such as saw milling and kiln drying. Improved marketing might offset some of the added short-term costs of more careful logging.

The survey

To help brainstorm better payment systems, I sent out a logger-payment survey to selected foresters and loggers. I did not intend to get a statistical sample and then choose a system based on majority preferences. Rather, the survey was designed to find out how individuals, who see themselves as careful stewards, deal with logger payment issues, and to ask them for ideas and advice.

I started by asking the respondents their status (loggers, foresters, or landowners), the scale of their operations, and the type of equipment they use. Those from different perspectives might favor different payment options.

I then mentioned two possibilities for who might sell the wood. The landowners might sell the wood to the loggers--who then cut it and sell it to the mills. Or the landowners can retain ownership of the wood and (with the help of foresters) sell the wood themselves. A landowner association would probably favor the second option.

With any payment method, there is an opportunity to encourage better practices with incentives. Landowners might want to encourage loggers to leave better residuals (and do less damage), do better bucking and sorting (to increase product value), or be more productive at cutting wood. Payment incentives include:

- job security (long-term contracts or right-of-first-refusal on next cut),
- monetary incentives for exceeding standards, or
- monetary disincentives for poor performance (i.e., penalties or loss of job).

I suggested that there are four basic situations that loggers and landowners might face--and each one might call for a different payment system:

1) high-value cut in easy logging conditions;
2) high-value cut in difficult logging conditions;
3) low-value cut in easy logging conditions; and
4) low-value cut in difficult logging conditions.

With higher-value wood, the logger would be doing a *revenue cut.* With lower-value wood, the logger would be doing more of an *investment cut.* With an investment cut, the landowner might make little or no money if the logger is adequately compensated. The landowner might still choose to pay for such a cut if the forester thinks the future stand will be more valuable. With low-value wood on a poor site with rough terrain, the landowner might decide that it is not worthwhile to cut at all.

I listed five possible payment methods, and for each one listed some considerations:

1) *By product and grade (standard piece rate).* While a piece-rate payment encourages productivity and good sorting, it can also be an incentive for high grading and stand damage. This method may mean more costs for supervision or for remediation. With low value stands, a logger might not get adequate compensation for his work. Because of pressure to get wood out fast, this payment system can lead to a higher accident rate.
2) *Straight rate by volume (or weight) regardless of grade.* A straight rate creates less incentive for highgrading. Because the logger is still paid by the piece, there is still an incentive to rush, which can lead to accidents. With a straight rate, the logger might not get adequate compensation cutting small wood in rough conditions. The logger and landowner might want to negotiate a different price depending on the wood and the conditions.
3) *Per time (hour, day, or week).* A wage per time would have to account for labor and equipment. With a payment by time, there is less incentive for highgrading and accidents, but there is also less incentive for productivity and proper bucking and sorting. Payment by time assures that the logger is adequately compensated regardless of the size of the wood and the condition of the terrain.
4) *Per area (by the acre or by the lot).* This system is often used for precommercial thinning. Payment would have to be based on the average size of trees, the stocking, and the difficulty of cutting. This system, like other piece-rate systems, creates an incentive to rush and may require more supervision.
5) *By formula.* In Scandinavia, foresters have computer software that takes into account stocking, percent removal, average size of trees, difficult trees, slope, roughness of terrain, yarding distance, and performance requirements. While this sounds complicated, the forester can type in the numbers for these variables in a matter of minutes and come up with a per-cord payment system. The trick is to have an accurate formula based on real data about performance in all these situations. Such a computer program has not been created for Maine, however.

I asked the respondents to choose which payment system (with incentive modifications if desired) would be most appropriate for the four logging situations. Finally, I left space where respondents could leave any comments. These comments, based on experience, were the most important part of the survey.

Results of the survey

The survey was sent to a small group of foresters and loggers who have experience doing higher-quality management. I was pleased to get nearly a dozen responses. The respondents included loggers, landowners, and foresters. Some were all three. They used all types of equipment ranging from horses to single-grip mechanized harvesters. They worked on small woodlots and ownerships in the thousands of acres.

Few of the respondents had experience with a formal incentive system, but most had recommendations ranging from logger performance bonds, long-term contracts, fines, and monetary incentives.

Except for payment by area, every payment system got an endorsement from some respondent for some situations. Managers of large areas (who could offer employment security) preferred straight payment by volume or weight, regardless of species or grade. Many of the other respondents suggested a variety of payment methods, depending on the wood and logging conditions.

Interpretation of results

This questionnaire was not intended as a scientific poll--the point was to get ideas and to learn from others' experiences. While I was hoping to get strong guidance towards the "best" payment system, instead, the variety of responses led me to a different conclusion: the system is not as important as the result. When the logger and landowner come together to negotiate a price, the landowner needs to make sure that the logger does a high-quality job at a reasonable price. The logger needs to make sure he makes a living--regardless of what is cut and the conditions of the forest.

It is possible to tinker with all the methods to ensure the desired results:

1) Landowners can accept reduced stumpage payments to take into account the logging conditions and the extra efforts needed to reduce damage.
2) The straight-rate per volume could be modified based on logging costs and conditions.
3) Payment by hour could be modified with financial incentives and disincentives to encourage quality, productivity, and value.
4) Payment by area could be modified depending on the conditions and the quality of performance.
5) Payment formulas could be devised that assure that loggers can earn a living regardless of conditions.

This conclusion leads to others:

- Loggers need to know their costs under different logging conditions.
- Foresters need to accurately describe the wood and logging conditions. These should be a standard part of the cruise.
- Loggers, foresters, and landowners need to negotiate a payment method based on the above information.
- Regardless of the payment method, the results should be similar - otherwise someone will not get a fair deal.

For a landowner association, it would be desirable to eventually come up with a single method that is less confusing. The association would benefit, however, even if multiple systems are used if landowners, loggers, and foresters keep careful records of costs, benefits, and general results. Such records would be crucial to a logger referral service. A logger with a good record and satisfied customers will be in demand and can get a better price.

Ultimately, all methods need to ensure that the logger makes a living wage and can pay for his machinery (or horse). If this is not the case, loggers will avoid low impact logging contracts. The simplest method to achieve that result would be to pay for loggers and machinery by time with clear job requirements and incentives for excellent performance. Even with this system, the logger has to be able to calculate how much the machinery is worth per hour (or day).

Appendix III of this book has a logging cost calculater work sheet to help loggers figure their costs given their type of equipment, the type of wood to cut, and the logging conditions. This work sheet can help enable clear negotiations with landowners, regardless of payment methods.

With low-impact forestry, initial cuts are often of low-value wood, leaving the best quality trees behind to grow and fill the forest. Low-value wood often goes into chips to be turned into commodities, such as pulp. Mills, through a variety of techniques, have been able to keep their purchase price for raw commodity wood low. Some landowners are overcutting now, flooding the market. The pulp market is depressed globally, due to overcapacity. In addition, the global market can, through artificially cheap shipping, supply chips from places like South America, to compete with Maine wood. Low prices mean there is little money to be shared by landowners, loggers, and foresters involved in managing such stands. This has created tensions when loggers or landowners feel shortchanged for their efforts.

When the wood is cheap, what produces it becomes cheap as well leading to cheapened forests, cheapened workers, and cheapened communities. A full cost accounting would lead to prices reflecting the cost of producing the product in a more sustainable way and giving adequate compensation to the producers. The economics of low-impact forestry will improve as the economic system is reformed.

Selected comments

The following are some of the comments from respondents:

- "Performance incentives would be a very positive addition to the equation. A payment system that takes into account all the variables involved in a harvest would be a sound point from which to negotiate a contract between logger and landowner. The gross price for a specific volume of wood (MBF, M lb, cords, etc.) is information that the landowner, logger, and forester need to share openly. From this mill delivered price, deductions are made by the parties involved based on mutually agreed on debits and credits from within a payment formula." (logger)

- "Have owners compete to hire the best logging crew--set a true market price for a logger." (forester)

- "Pay $15/hour base plus XX/MBF to encourage production. Give 10% extra for minimizing stand damage" (forester)

- An incentive for loggers is "reasonable assurance of an average production over the operating year. Workers are paid salary." (forester)

- Payment by time "sounds like an employer/employee relationship. Therefore worker's compensation insurance is required under Maine law." (forester)

- "I have tried to explain to landowners the hidden costs of harvesting. A higher bid does not guarantee a good job. That is why it is important to view past jobs of contractors or to get references." (forester)

- "I think #5 (formula) is a good idea if and only if the landowner is educated on exactly what they are getting for their investment in thc process." (logger)

- "Stumpage prices as a % of mill delivered would encourage utilization. Hourly pay to loggers would allow them the time to do a good job. An additional incentive based on some measurable unit could be added. e.g.. a (?)% bonus if (?) % or less of the residual stand has damage smaller than a credit card." (forester)

- "I used to get good results for owners and loggers when the logger was hired to land product roadside and was paid by owner (usually a forester) by the day.... I have used bonus incentives and penalties in many of my contracts over the years... (one) incentive was land base (opportunity for long-term work). Stumpage prices were negotiated every year and were based on a percentage of mill price. No bonus." (forester)

- "There are several ways to get to the desired result. None will work without honesty, trust, and clear and almost continuous communication. Much easier said than done. I have numerous horror stories to tell..." (forester)

- "How do you make 'stewardship contracts' (right of first refusal for other cuts) legally binding to control forgetful landowners?" (forester)

- "Try to get logger to actively participate in negotiation (how much will it cost them to cut as marked, as specified in advance) then add bonus/incentives for smart marketing (exceeding average price/MBF for example)." (forester)

- "I require loggers to control/limit residual damage to less than 10%--that is a contract requirement just like putting water bars in. Good work means job security and a returned performance bond. Bonuses are only paid for active marketing; then on a % basis. Performance bond would balance logger's short-term interest with landowner's long-term interest." (forester)

*****

11. Patient Money: the Economics of Low-Impact Forestry

If low-impact forestry has clear advantages for minimizing stand damage, maintaining wildlife habitat, protecting aesthetic resources, and enhancing recreational aspects of a forest, why isn't it the forestry norm? Many people might conclude that LIF must be uneconomic--the costs probably outweigh the benefits, otherwise more people would be doing it. If LIF makes economic sense, there would be few excuses to not practice it.

Low-impact forestry does *make economic sense within an ecological/social perspective and the assumptions that go along with that perspective. This perspective assumes that economics is embedded in nature and society and not the other way around. Not all forest landowners reading this chapter will immediately switch over, however. Forestry economics is complicated and not all landowners operate within the same perspective.*

LIF assumptions. Low-impact forestry economics operates from the following assumptions:

- Look at *Total Value* (removals plus residuals), not just removals. From this perspective, damage to residuals is considered a *cost*.
- Consider both long-term and short-term, costs and benefits. We assume there will be a future. Analysis of sustainable forestry should consider impacts over generations. What maximizes returns over the short term may be doing so at the expense of the long term.
- Look at impacts to all the players (landowners, loggers, and the local community), not just one at the expense of others. While it is possible to get higher returns for landowners by exploiting labor, for example, doing so hurts the whole economic system (the community) to which the landowner belongs. From the holistic perspective, such benefits to a part are not a benefit to the whole.
- Avoid externalizing costs to others or to future generations. Damage to productivity, water quality, soils, residual trees, aesthetics, or property values should be considered costs--even if an exact dollar value can not be easily attributed to them and even if "someone else" pays the costs.
- Do not confuse income with capital depletion.

Income or Capital Depletion?

The confusion over income and capital depletion comes from ignoring the concept of *total value*. Consider the following analogy with financial investments. If you have $10,000 in a fund that earns 10% a year, you could remove $1,000 a year and still have your principal undiminished each year. But suppose you remove $1,000 one year, $1,500 the next year, $2,000 the next year, etc.. You might get excited and brag to your friends that your "income" from your investment is growing every year. Unfortunately, after 5 years there will be no "income" because there will be no principal. If, over time, what goes out of the system is greater than that which goes in, the system is not sustainable.

Forests have natural capital that affects growth and stability such as:

- the stocking, size, and quality of standing trees;
- soil fertility factors--including species and processes that affect nutrient availability;
- species and processes that increase resistance to catastrophic disturbances;
- biological legacies that persist through disturbances and improve resilience;
- genetic adaptations of trees and other species to the site and to potential change over time.

To the extent that this natural capital is diminished at a rate faster than it can be replenished or to the extent that its ability to function fully is compromised, the forest principal has declined and true sustainable income will be less. Such damage, therefore, is a cost to be avoided.

Landowner objectives. Foresters and loggers tend to meet landowner objectives as a first priority, even if these objectives conflict with sound silviculture or sound ecology. Not all landowners share the LIF goals as their prime objectives. Landowner objectives and landowner economic perspectives vary widely due to such factors as:

- *Type of landowner.* Public ownership may have requirements for "multiple use." Contractor-owners may be more concerned with cutting enough wood to meet payments on equipment or land than with managing for the long run. Some small woodlot owners may value the land more highly as an aesthetic neighborhood buffer than as a major source of income.
- *Size of ownership.* Small landowners who want a steady income avoid clearcutting. Clearcutting gives a big pulse of income (and income taxes) and then expenses for early-stand management followed by a lifetime of no income (but continued property taxes). Large landowners, in contrast, can justify managing in blocks and balancing early-stand expenses with income from final cuts elsewhere. Some landowners are so large that they can distort local markets in their favor. These types of landowners can, for example, leverage wage levels below what would be a free-market price--because these landowners may be "the only game in town."
- *Degree of vertical integration.* Some industrial landowners, who do not see their woodlands as a "profit center," can justify "selling" wood, which could have become sawlogs, as pulpwood to their own mills--if this floods the market enough to keep purchase prices low. Maximizing income for the woods division may not be as important as assuring a cheap, stable supply for the mill.
- *Location of headquarters.* It makes a difference if the landowner is absentee or lives on the land. Resident landowners are more apt to be more concerned over community costs that they will have to live with. Local owners tend to spend more profits locally, enhancing local communities.
- *Location of timberlands in relation to markets and labor.* Distance from markets can affect stumpage, mill-delivered prices, and trucking. It also makes a difference if the labor is migratory or lives in the same community as the land. Local labor will tend to spend more wage money locally.
- *Presence and availability of loans, subsidies, taxes, or tax breaks.* When land is purchased with large short-term loans, the perspective of the landowner on management is different from those whose land was bought generations ago. Big short-term returns are required. Management decisions can change when government "assistance" is available. Clearcuts, for example, become much more viable if someone else pays for the required early-stand management expenses.

  When taxes are low enough, holding heavily-cut land that has low productivity becomes less of a burden. Low taxes allow a landowner to cut heavily now, and take some time to find a buyer for the cut-over land.

  Subsidies can make practices that normally would be uneconomic more cost-effective. This could shift investments into less efficient directions.
- *Regulations (or lack of them).* Regulations might restrain certain types of cutting in certain types of areas. Landowners are not supposed to clearcut riparian zones, deer yards, or high-altitude zones on mountains. On the other hand, to the extent that regulations allow

abusive cutting, subdivisions, and sales in areas with high land values--even without mature forests on the ground--landowners who want to maximize their short-term returns can be amply rewarded. Such short-term gains can be more enticing to some landowners than the long-term benefits of LIF.

Logging economics. Even when landowner objectives are similar, logging economics can vary widely due to the following factors:

- *Type of loggers.* It makes a difference whether the loggers are large contractors, small contractors (who do the cutting themselves), employees of the landowner, or the landowners themselves. These differences will be reflected in differences in both costs (such as workers compensation) and benefits. Larger contractors, for example, may be able to secure higher wood prices, but they may also have heavy debts for equipment, compelling heavy cutting. Owner-cutters can keep a higher proportion of mill-delivered prices and can more easily justify more careful practices.
- *Type of forest.* The stand type, soil type, stocking, tree size and quality, scale of cut, slope, season, and presence of sensitive areas (such as water bodies or deer yards) can all have major impacts on costs and benefits from a logging operation. Some stands, due to their poor quality, remote location, or difficult terrain, may not be worth cutting at all. While some landowners with a long-term perspective might have little trouble justifying low-impact forestry on the whole ownership (for example, land trusts, institutional owners, families with roots), even owners with shorter horizons may find it justifiable to use low-impact methods in sensitive areas, such as riparian zones, or near recreational areas.
- *Type of equipment.* The economics of logging equipment depends on whether the machines are used or new, and the appropriateness for the site, the size, and type of cut. Some machines that can do a good low-impact cut in some stands, for example, may be inappropriate for others because the machine is underpowered to haul big wood, or because yarding distances are too long. The economics of the equipment also depend a great deal on the operator and planning. A good cable skidder operator might be able to do a relatively-low-impact cut in an economic fashion. A poor operator could mangle the forest and rut up the soil.
- *Type of cut.* The same machine might be appropriate for one type of cut but inappropriate for another. Some large machines with high-flotation tires might be adequate for a heavy cut, but be too wide for a light cut that requires narrow trails that allow crown closure.
- *Type of market.* The economics of logging improve as the value of the wood goes up. The same wood may have a significantly different value depending on the type and location of the market. In some cases, the market value of the wood is so low that the costs of logging and trucking are higher than the mill-delivered price. In such a case it may make more sense for the logger to leave the wood in the forest to rot.

  Even for the same product, market prices can vary widely over the years, or even within one year. Shortfalls, oversupplies, and events far away can all cause dramatic swings in prices. Timing of the cut is thus an important factor in the economics of logging. For some commodities, mills that have a dominating influence over the market can get away with paying much less than what would be a true market value, which hurts both landowners and logger.
- *Cheating.* Some loggers improve their economic prospects by stealing stumpage, lying to landowners about how much wood is cut, or underpaying workers (pay for "equipment,"

rather than labor, for example, to avoid insurance or workers' comp payments). While cheating may help the prospects of the logger, it obviously does not benefit the landowner.

Each one of these factors for both landowners and loggers can cause values to vary so widely that a concise economic analysis to three decimal places is a mockery. Such variability defies an economist's ability to do comparisons with exact numbers. If one multiplies the range of variability, the result qualifies as an example of chaos.

A barrier to low-impact logging. On a good site with high-value wood and with a skilled operator, a low-impact logging operation can easily pay its way, leaving both landowner and logger happy. Such stands, unfortunately, are not the rule. Too often the stand to be worked has been repeatedly highgraded and is filled with poorly-formed and damaged trees.

The initial cut in such a stand might consist of removing low volumes of low-value wood--leaving the best wood to grow. This type of cutting takes considerable skill if the logger is to avoid damaging the residual trees and soil. It may even take specialized equipment that is not as productive as conventional equipment. The result can be that the cost of logging goes up while the value of the cut doesn't. Only so much money is available in a cord of wood. Few loggers are motivated to work harder to make less money. Few landowners are willing to have their land logged and see little money in return. Both landowners and loggers have expenses to pay. Feeling good about a cut, unfortunately, is not sufficient for paying bills. Most people prefer cash.

The barrier of low immediate cash leads some landowners to dismiss low-impact logging as "too expensive." These landowners might reason that since there are so many trees out there, it shouldn't matter if a few get damaged during logging. Some loggers might conclude that they cannot afford to do low-impact practices and be competitive at paying stumpage. Other loggers will get the bid.

In 1997, using standard accounting formulae and actual data from contractors, I compared the cost per cord of a low-impact system operating on woodlots with a mechanized system operating on industrial lands. The low-impact system used a chainsaw and forwarder equipped with a radio-controlled winch. The mechanized system used one feller buncher, two grapple skidders, a crane, and a delimber (at half time). The costs (including labor) were $45 a cord for the low-impact operation and $38 per cord for the industrial operation. Thus the low-impact system cost 18% more per cord.

Admittedly, this is an unfair comparison, because the objectives, cutting systems, and sites were different. The mechanized system is designed to cut as much wood as fast as possible. It would be very difficult for the mechanized system to leave as much wood with as little damage as the small forwarder with the radio-controlled winch. The low-impact system was also inefficient because it was a one-man operation--the machinery stood idle while the logger cut and limbed the wood. Even if the costs were more competitive for the low-impact logger, the revenues from the cut would not be. The industrial system was cutting the highest value wood, leaving little or nothing of value behind. The low-impact system was cutting mostly lower-value wood; it cost more, but had lower immediate returns in revenue.

What factors might justify low-impact forestry when immediate returns seem meager?

Immediate benefits. Although the advantages of low-impact approaches stand out most clearly in the long term, there are significant short-term benefits as well:

- *Higher residual value.* If one figures costs and benefits based on *total value* (removals plus residuals), then low-impact logging has some advantages. First, having narrower and fewer trails allows the landowner to retain more crop trees (trees of high potential value). A mechanical operation with trails every 40 feet might have to remove 25% of the quality trees (before they are ready to be cut)--just to make way for the machinery. Second, the low-impact cut damages fewer trees, leaving more a more valuable residual stand.
- *Higher property value.* If property values are determined by raw-land value plus timber value, then the property value of the land would be higher after a low-impact cut than after a more standard operation. Not only is the timber value higher, but the aesthetic and recreational values are higher as well. High-grading operations can not only lower the property values of the land cut, they can also have a shadow impact over adjacent lands. Poorly-cut lands can be visually distressing and can lower the value of neighboring properties. To large, absentee landowners, unfortunately, this may not be an issue.
- *Higher wildlife value.* Although wildlife values are best measured with biological, rather than economic criteria, retaining large trees, more shade and vertical structure, and more interior forest is a clear advantage. In some parts of the state, such habitats are in short supply.
- *Better quality soil and water.* Low-impact cuts lead to less compaction and rutting of the soil, more shade to the soil, and less chance of silting and warming of streams.
- *More jobs.* In the example I gave comparing the logging costs of two cutting systems, the major factor for higher costs of the low-impact system was labor. Labor was 60% of the cost of the low-impact system, but only 25% of cost of the mechanized system.

While labor to a logging contractor might appear as an unwanted cost, to a community it is a benefit. Money paid to labor multiplies in the local community more than money paid for machinery and fuel. Much of the money for machinery goes to out-of-state equipment manufacturers, banks, and oil companies. Money paid to labor leads to more local spending on food, housing, entertainment, and other goods and services, thus supporting more local jobs. Even excluding these multiplier effects, for the same volume of wood cut, the low-impact system analyzed would employ three times as many loggers as the mechanized system. For horse logging, the difference would be even greater.

Long-term benefits. Some people mistakenly see forestry as an investment which, at best, can only give low returns. Forests grow slowly--sometimes only two or three percent a year--and thus one can only get a two or three percent return. When one factors in inflation, then the outlook is not quite so bleak--one can get two or three percent return above inflation. For example, if inflation is four percent, one can get a seven percent return--which is, at least, better than putting the money in a savings account.

However, if long-term economic returns are important, five factors should be considered (plus a few others that are harder to measure with dollar values):

1) Growth and/or volume/acre/year;
2) Species mix;
3) Product mix (pulp, logs, and grades);
4) Market value changes compared to inflation;
5) Risk from insects, disease, or wind;

plus social and biological considerations.

The combination of the five factors (plus) can lead to significant advantages of low-impact forestry over more conventional practices with shorter-term horizons. Indeed, when one considers these factors, it is hard to justify *not* moving toward lower-impact forestry.

Growth. How the forest is cut can affect the volume growth per year in a number of ways:

- With LIF, more land produces trees because less land is taken out by trails, yards, and roads. Less land in trails means more residual volume on which to have growth and more crop trees per acre. A mechanized system might remove 25% of the potential crop trees just to make trails, while a low-impact system might remove 8-10%.

Excessive feller buncher trails lower future productivity. Aerial photo by Mitch Lansky.

- LIF leaves less residual damage of trees and soil. When tree trunks or roots are damaged, the tree has to expend energy compartmentalizing the wound--energy that could have gone to growth. Tree damage also means an increased percentage of growth is going on lower value products. Hurting soil structure can lead to less available air, nutrients, and water to the tree roots and can impede root growth physically, as well.
- LIF favors leaving dominant, vigorous trees. Variations on diameter-limit cuts (where the biggest trees are cut, leaving the smaller ones behind) can remove the dominant trees and leave suppressed trees with smaller crowns and root systems, lowering growth rates as well as putting growth on lower-valued trees.
- Large trees are more efficient than small trees at producing stem per unit of leaves.
- Shade tolerant trees can grow under the shade of larger trees, leading to better use of the growing space than in even-aged stands.
- With shade tolerant trees already present in the understory, harvested trees are replaced faster in uneven-aged stands with less cost than in even-aged stands.

- With even-aged stands, one long rotation produces more volume at higher quality and value than two short rotations.

Species mix. LIF favors retaining longer-lived species that are suited to the site. These species are capable of growing to larger diameters and high-valued products. Shifting the species mix can lead to major differences in the value of the growth of the stand. For the same sized tree, for example, a rock maple sawlog might be worth twice as much as a red maple sawlog. Red spruce can live longer, and grow to provide more valuable products than balsam fir.

Product mix. LIF not only encourages the growth of larger trees, it also discourages tree damage that might lower the grade of a log. Dramatic increases in value occur as trees grow to different products and grades. In 1998, the average cord of rock maple veneer was worth 70 times the value of a cord of rock maple biomass.

Favoring crop trees does not lead to the total elimination of poor quality trees, but it gradually shifts the percentages of products so that more of the growth is of higher value. Trees of lower economic value still have important silvicultural value (for shade and windfirmness) and ecological value (for habitat).

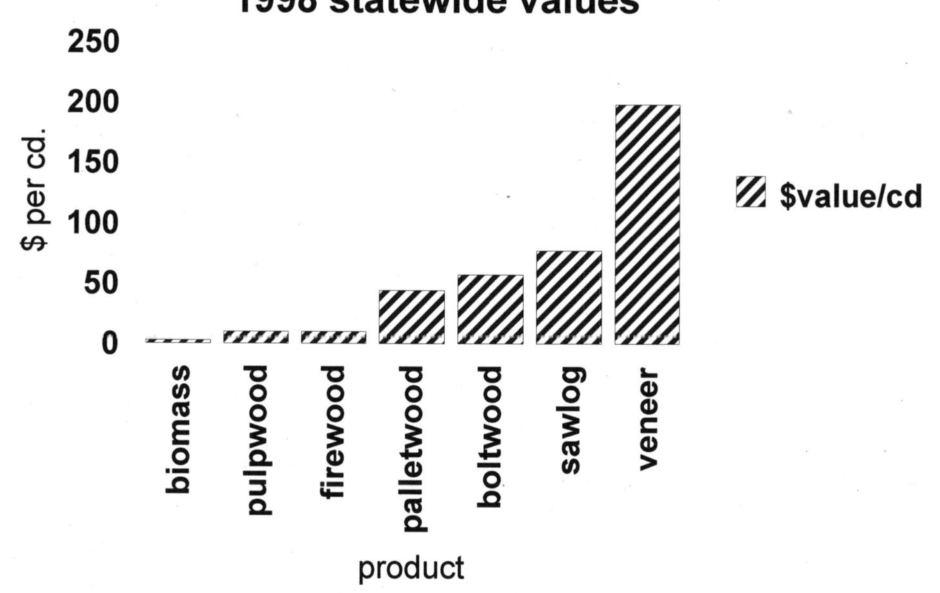

Impact of product mix on future value of stand

The following charts (based on data from USFS guide to hardwood silviculture) illustrate the benefits of managing for improved product mix. Product mix "A" has the lowest percentage of high-quality timber; "D" has the highest. In the following graph, line ABCD shows what happens to returns when product mix is improved over time from "A" to "D."

Assumed percentages of sawtimber volume

Product	Product distribution			
	A	B	C	D
Veneer	2	4	6	8
Sawlogs				
High Quality	3	6	9	12
Medium Quality	40	45	50	55
Low Quality	15	15	15	15
Pallet Stock	40	30	20	10

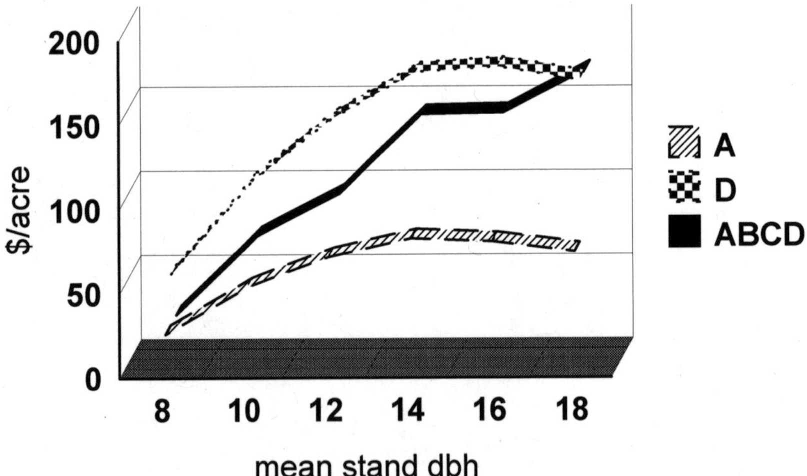

Market value changes. An examination of stumpage prices in Maine since 1959 shows that inflation-adjusted prices for commodities (such as pulp) hovered around inflation, but sawlogs showed modest increases in value above inflation, and veneer showed dramatic increases in value above inflation. Indeed, during the 1990s, rock maple veneer had an average increase in value above inflation of around 20%! Even if a rock maple veneer tree had ceased to grow, it could have been a good investment *not* to cut it (assuming its quality did not decline).

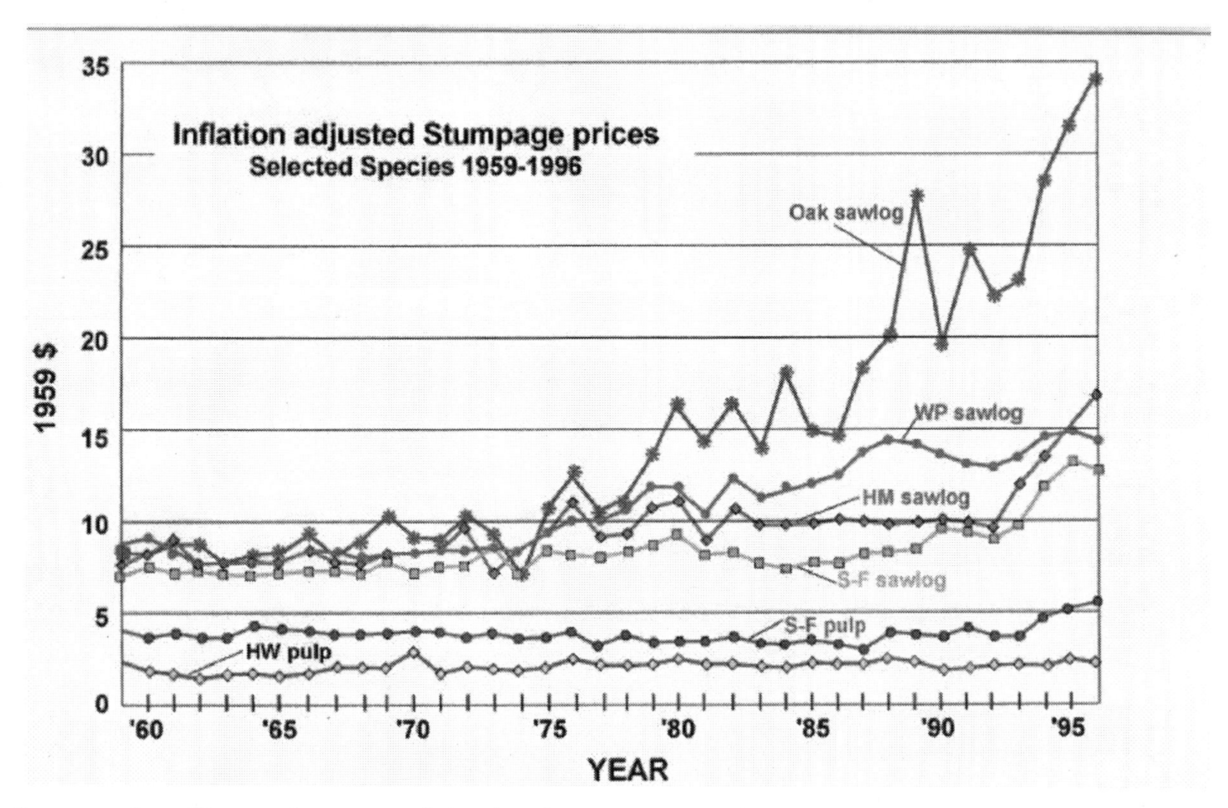

Past trends will not always work in the future. New technologies of composites and laminates may change consumer preferences. High quality timber, however, is likely to continue to have favorable demand. Combining tree growth with growth in value above inflation makes protecting these high quality trees from damage a no-brainer.

Risk. All investments have a certain degree of risk. Junk bonds, for example, paid high returns, but they also had high risk--there was no guarantee that the investor would see those returns. Forestry investments also have a degree of risk from fire, wind, ice, insects, or disease. LIF lowers these risks compared to more standard approaches.

Well-stocked stands, for example, are less subject to windthrow. Favoring dominant trees for retention means leaving trees with stronger root systems that lower the chance for windthrow. Suppressed trees that get "released" when dominant trees are removed in a diameter-limit cut are more susceptible to windthrow. Long-lived, vigorous trees are also less likely to be killed by insects or disease. Trees wounded by logging have a higher chance of being invaded by insects or fungi.

Because LIF tries to leave behind a forest with a fuller range of habitats, including large trees and even dead-standing trees, the stand is more likely to have a fuller range of predators and parasites of potential pest species. For example, certain warbler species (Cape May, bay-breasted, and Blackburnian) that prefer large spruce trees for habitat can be an effective check on early stages of spruce budworm outbreaks. Stands dominated by young fir (as might be the case after a clearcut or heavy highgrade operation) will likely not have these species. Even if trees die or fall over, the LIF trail system allows for effective salvage, so that the death of a valuable tree (that is not necessary for structure) need not be a financial loss.

Social considerations. In addition to more jobs per cord cut (mentioned under short-term benefits), LIF, because it grows higher-quality trees over the long term, can lead to increased stumpage revenues for landowners and increased opportunities for local sawmills. Improved aesthetics and recreational opportunities can also benefit the local community.

Biological considerations. The most valuable habitats are the hardest ones to grow. It is easy to create early successional habitat in a matter of days with a few big machines. In contrast, it may take a century to grow interior, late successional habitat. Growing big trees in well-stocked stands takes time. LIF favors retaining large trees (both live and dead) in well-stocked stands. Although LIF is no substitute for reserved forest, it compliments, rather than isolates, reserves, and therefore increases, rather than decreases, the value of reserves or wilderness.

But can you get there from here? "There" is a well-stocked stand of high-value wood. If "here" is a

Mel Ames proves you can get there from here

Mel and Betty Ames have been managing their woods in Atkinson for over 50 years. When Mel first started buying the land, in the 1940s, it had been cut over and only had around 15 cords to the acre. He now has around 600 acres of forest. After decades of cutting, most stands now have 20-30 cords to the acre, with some averaging 35-40 cords and other stands (with pine) having more than 50 cords to the acre. His growth rates are now a multiple of the state average. The proceeds from his cutting helped Mel and Betty raise eight children.

Mel got his forestry training at nearby Foxcroft Academy. The Academy was inspired by Maine forester Austin Cary to advocate sustained yield, rather than cut-and-run forestry. When he graduated in 1946, Mel started buying up woodlots and managing them based on what he learned, and as time went on, based on his experience.

Mel does not have rigid, long-term management plans. He cuts trees because there is a market and he wants to make money. But he has an inner sense about how to cut. He waits until trees have their highest values. He keeps stands well spaced for optimal growth. He ensures the stands are stocked with quality trees. He works with the species that are growing. If the stand is early successional, with a poplar overstory and a fir understory, that is what he grows. When he is cutting bigger trees, he sometimes thins adjacent saplings if they are crowded.

Because he works with succession and keeps full stocking, he is creating more late-successional habitat. He has pine martens living in his woodlot and birds that need interior forest.

Mel does not impose harvesting systems to stands. He cuts in a way that leaves a forest stocked with good quality trees. When he is done (usually he removes around 25-30% of the trees), you could label it "selection," "irregular shelterwood," or some other silvicultural system, but that "system" is a response to the stand conditions.

His woods technology began with horses, then jitterbugs, crawlers, bombadiers, a tracked skidder, and now a small skidder (John Deere 440). He does minimal damage to stands with cable winching and skidding the logs out. Sometimes he forwards shorter-length wood with a trailer. He and his son Russ are careful in the woods because it is obvious to them that damaging residual trees is destroying future value. They are now cutting high-quality sawlogs and veneer that Mel had tended over many decades. They can see the results. Mel jokingly says that well-managed forests can beat blue-chip stocks for returns.

stand with enough good wood to more than pay for the cuts, the answer is easy: "yes." Indeed, it would be poor judgment to damage such a stand with heavy-handed cutting practices.

If "here" is a stand that has been repeatedly highgraded and is dominated by low-quality wood, "there" will take a long time, and may require some "investment cuts" that pay poorly, if at all. The long wait and the need for "investment" cuts is the penalty to be paid for previous mismanagement. Those landowners who did the previous cuts probably thought that what they were doing was economically sound. But they were not accounting for the costs to those who had to make the next cut. They were doing forestry as if the future did not matter.

If "here" is a stand on poor soils, with difficult terrain, poor accessibility, and short-lived, low-value species, the answer can be: "no." The stand may not be worth managing for timber. It may be more valuable to protect the soil or water or for wildlife habitat.

Government policies. In some cases, government programs, such as those that subsidize timber stand improvement, can improve the economics of "investment cuts." Foresters are generally knowledgeable about the existence of such programs.

Government policies, however, can lead to "perverse subsidies" that hurt the economy in general as well as the forest. Policies such as direct payments, lax environmental standards, lax labor standards, allowance of market domination, or pork-barrel tax breaks, can, in some cases, distort investments towards inefficient directions. Artificially cheap commodities can lead to waste and overuse. For woodlot owners, low mill-delivered prices, due to domination of markets by a small handful of players (oligopsony), makes getting good returns on forest practices more difficult. Rather than remedy this inequality by subsidizing woodlot owners, a better remedy would be more fair prices, which would help out both woodlot owners and loggers.

Computer models. Computer models are useful tools in making decisions on what is manageable and what kind of returns one can expect. With some models, such as FIBER (from the US Forest Service) one can program in cruise data to see what the stocking, volume, and value of the stand are. One can then program in a management system and get predictions of future growth, species ratios, product mixes, and values.

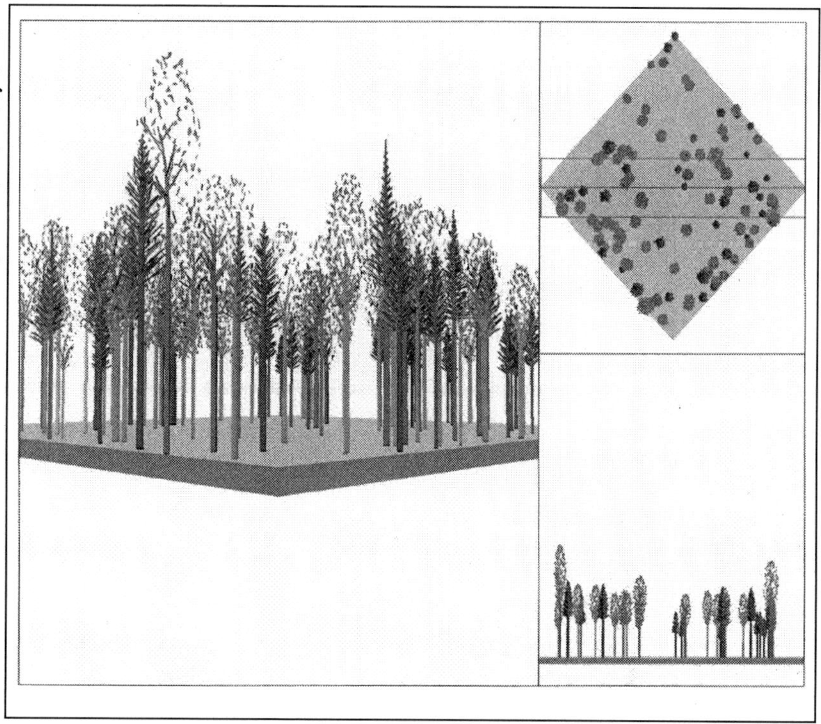

Graphics generated by FIBER FLEX computer model

Models do have limitations. They cannot predict:

- changes in the economy--including markets and prices;

- wind, ice, droughts, insect outbreaks, or climate change;
- the impacts of trail and yard distribution, or the impact of residual damage to future growth and quality;
- political shenanigans, wars, depressions, cartels, globalization, terrorism, or social turmoil...

  The world continues in the simplified computer model without serious disruptions--which is abnormal in the long run. But the models can be a useful tool, nonetheless, in visualizing possible consequences of various management approaches to a given stand and deciding if management makes silvicultural or economic sense.

Forestry associations. Forestry associations[1] can improve the economics of woodlot management a number of ways:

- Educate landowners, loggers, and foresters-- through literature, workshops, demonstrations, and trainings--to do more efficient management practices and marketing;
- Give referrals for good quality foresters and loggers;
- Help with pre-and post-logging assessments to determine level of stand damage;
- Set up concentration yards so that wood can be sold in more viable quantities;
- Establish cooperatively-run mills so that landowners can make money from kiln-dried lumber, rather than just stumpage;
- Establish logger associations so that loggers can save money using combined equipment, improved marketing, and improved techniques;
- Find higher-paying markets--including marketing certified wood.

Other considerations. The economics of logging are affected by more than just the management and the marketing.[2] The answers to the following questions can sometimes make a big difference in economic outcomes:

- How are deeds set up?
- How is forestry financed?
- What accounting procedures are used (such as net Present Value, Internal Rate of Return, Managed Forest Value, dollars per acre per year, etc.)?
- What discount rate is used? (if the investment is multi-generational, does it make sense to use a high discount rate to determine present value?)
- How are contracts set up?
- How are taxes paid?
- What has been done for effective estate planning?

Sustainable? One low-impact cut does not mean forestry is sustainable. It does keep options open for forestry to be sustainable, but there is no guarantee that the land will not be liquidated later. This issue can be addressed in a number of ways:[3]

- deed restrictions,
- easements,

---

[1] See section on establishing woodlot owners' associations

[2] See section on legal aspects of owning woodlands.

[3] See *Conservation Options: a guide for Maine landowners*, by Maine Coast Heritage Trust, Northeast Harbor, ME

- land trusts,
- forest banks,
- public land,
- public policy (research, demonstration, outreach, education, tax policy, regulations, etc.), and
- a strong conservation ethic.

A strong conservation ethic that is passed on through the generations is the most important item on the list. What protections one lawyer can put together, another can take apart. It takes a strong conservation ethic to prevent the current generation from taking the gold mine and leaving future generations the shaft.

Indeed, the key to the economics of low-impact forestry is that the benefits of actions taken today are going to more than this generation. As Wendell Berry wrote in his essay, "Conserving Forest Communities":

"The ideal of the industrial economy is to shorten as much as possible the interval separating investment and payoff; it wants to make things fast, especially money. But even the slightest acquaintance with the vital statistics of trees places us in another kind of world. A forest makes things slowly; a good forest economy would therefore be a patient economy. It would be an unselfish one, for good foresters must always look towards harvests that they will not live to reap."

*****

New Economic Analysis Tool for Woodlot Owner

Brooks Mills, a Maine woodlot manager, and Neil Lamson, forester for the State of Vermont, have created a spreadsheet program that can help woodlot owners calculate the economic benefits of growing quality timber.[1] The Brooks and Neil Tree Investment Chart (BAN-TIC)[2] uses information about grade (based on number of clear faces), volume, growth rates, and prices to project volumes, value, and rates of return over time of individual trees 10-30 inches DBH. With the program, users can calculate financial maturity of a given tree as well as the value of investing in management to improve growth rates or quality. It can also be used to calculate stand values.

The program demonstrates why it is a good idea to identify "crop trees," trees worthy of special attention and culture. It also demonstrates why landowners could benefit from growing these crop trees to larger diameters, rather than cut trees as soon as they reach sawlog size. The program assumes two 10-foot logs per tree. As the diameter increases from pulpwood to sawlog, and from sawlog to veneer sizes, the tree's values increase dramatically. Given a better grade of sawlog, annual returns from letting a tree grow can outpace most conservative investments even with trees up to 26 inches in diameter.

Brooks emphasizes that the really important thing that landowners can do is "to mark and number these high-value trees when they are 12-14 inches dbh and find out how they are doing and follow their development." In some cases, thinnings can help maintain tree vigor and lead to higher rates of increase in value. These measurements can also help the landowner spot when tree vigor is low and help in making the decision to cut before the tree goes down in value. Poor vigor in maple or ash, for example, can cause the dark heartwood to expand and lower the grade from veneer to sawlog.

The benefits of growing larger, higher-quality trees are even greater than shown in the chart for a number of reasons:

---

[1] For an example of the spreadsheet, see appendix IV

[2] For a copy call SWOAM office at 1-877-467-9626

- For high quality timber, mill-delivered and stumpage prices have been increasing at a rate faster than inflation. This increase is not accounted for in the program.
- Larger trees can sometimes produce longer high-quality logs.
- Rate of return is not necessarily the best way to calculate the value of letting trees grow. Instead of looking at the rate of return (in percentages), a landowner can look at the dollar value increase per year per tree (or per acre). Annual increase in dollar value for 28 inch trees growing 2 inches is greater than that for 14 inch trees, even though the rate of return is far greater for the smaller trees. For example, a tree yielding two 10-foot logs with four good faces growing from 14 to 16 inches yields a 36% annual rate of return, but going from 28 to 30 inches yields only a 2.7% annual return given a growth rate of 7 years to increase 2 inches in diameter. The dollar value change, however, is $122 dollars for the 14 to16-inch increase, but $255 dollars for the 28 to 30-inch shift.

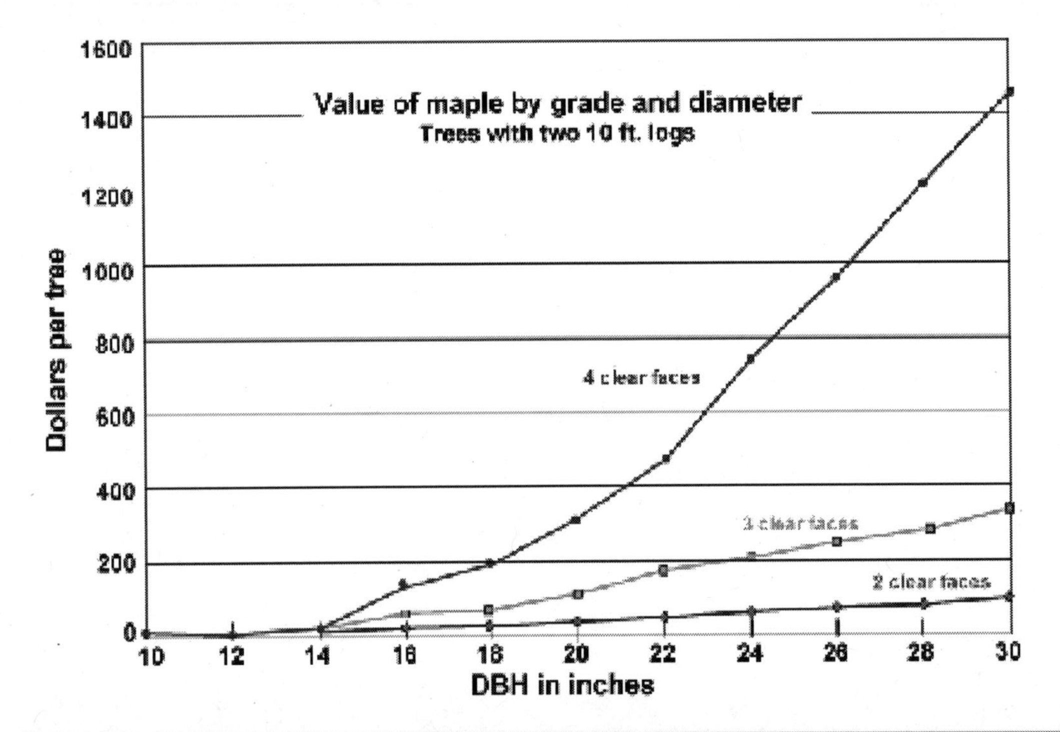

An individual tree is not an investment. You can not buy a 10-inch tree to plant on your land. It takes many decades to replace such a tree. Since larger trees put on more value per year than smaller-diameter trees, it does not make much sense to cut the trees when small and grow another small tree to take its place. It makes even less sense to damage a potential crop tree, lowering its growth rate or lowering its grade, which is why low-impact logging is so important for growing quality over the long term.

While landowners can use such charts to calculate the financial maturity of a tree, landowners might also be concerned with the biological maturity of a tree. The biological value also tends to increase with size. Growing trees past "financial maturity" can still yield better returns than cutting trees well before their prime, unless holding the trees longer would lead to rot or staining that would lower the grade.

12. Legal Aspects, Contracts, and Land Protection

This chapter looks at some forestry issues that go beyond just cutting trees, such as contracts, deeds, taxes, and long-term protection. This chapter points to several resources woodlot owners can use to explore these issues and gives an example of how one family is putting various approaches together to ensure management and protection for the long term.

Legal Aspects

Thom McEvoy, an associate professor and extension forester at the University of Vermont, has written a book that helps woodlot owners realize what they need to know, and where to go for help concerning legal issues surrounding woodlot management. McEvoy's book, *Legal Aspects of Owning and Managing Woodlands*, discusses many important questions for forest landowners:

- What are the rights of woodlot owners?
- What are the types of deeds, and which ones should you have?
- What are the various methods of ownership, and which ones make the most sense in a given situation?
- How do you locate and maintain boundaries?
- How do you do forest management planning?
- What should you look for in contracts with foresters and loggers?
- What do you need to know about forest taxation?
- How do you plan your estate so that after you die, the woodlot is cared for according to your wishes, and your heirs are not forced to cut or sell due to high estate taxes?
- How do you settle disputes regarding your woodlot.

Woodlot owners who ignore these questions do so at their own peril. Knowledge of these subjects is key to ensuring long-term management continuity and assuring optimal financial returns. One can make excellent forestry decisions, but poor tax decisions, for example, and lose money when you should be gaining. One can take exceptional care of a woodlot for decades, and then, after your death, the woodlot might get liquidated due to poor planning.

The chapter on estate planning helps landowners to confront their own mortality and to think in the long-term. A landowner who wants to see the forest cared for over many generations has to decide who, after his or her death, gets the land, and under what conditions. This chapter also discusses options such as easements or land trusts.

The book also has a chapter on professional ethics. Certain actions, such as heavy high-grading, may be legal, but they are unethical for foresters to recommend. Likewise, one-sided transactions can tread in the gray areas of ethics. When wood buyers give estimates of wood volume, and the woodlot owner has no way to verify these numbers, ethical problems arise, especially if the landowner is paying based on the estimates, rather than on actual removals. Foresters who get paid on a percentage of what gets cut can have an economic interest to cut more, regardless of the conditions. Ethical tensions are common over the conflict between choosing what is right ecologically and what makes money in the short term.

Some of the more technical sections on taxation, estate planning, or deeds, are more comprehensible to professionals, than to the lay person. The book, in trying to explain these complexities, sends the strong clue that there are times when the best path is to hire a professional forester, lawyer, or estate planner.

LIF Contract Contents

The following are items to be included in a contract between the logger and the landowner. Each contract needs to be tailored to fit the specific forestry situation, the desires of the landowner, and the set-up of the contractor. For low-impact forestry, it helps to be clear about standards the contractor should meet and what compensation the landowner will pay the contractor to meet those standards (or what penalties if the contractor does not meet the standards). The items below can be inserted into the "model contract" in appendix V.

- ✓ The location of wood to be cut (and other management activities).
    - Who is responsible for marking boundaries and providing access?
- ✓ Who owns wood and who will sell it to mills.
    - Contractor buys from landowner and sells to mill?
    - Landowner retains ownership of wood and sells to mill?
    - Forester acts as agent of landowner to sell to mill?
- ✓ How to pay and what for.
    - Lump sum (agree to pay lump sum regardless of volume cut)?
    - Mill tally (pay per unit based on mill scale)--what is rate per cord per species and product?
    - Share basis (pay a percentage of revenues from cut, such as 20%)?
    - Services rendered (per hour, day, week? This rate should account for equipment costs and labor costs on a time basis)?
- ✓ Role of forester as agent of the landowner (for example, monitoring contract compliance, or even selling wood).
- ✓ Indemnification clause to limit liability. "Save and hold harmless" the seller.
    - Contractor has filed "predetermination" form.
    - Require workers' compensation and liability insurance of the buyer.
- ✓ Expiration date for contract.
- ✓ Agree to settle disputes through arbitration.
- ✓ Performance bond (or money held in escrow) to ensure that forest assets are not damaged (if such assets are considered at risk).
- ✓ Expected outcome for trails, slash, limit of damage, and following BMPs (LIF standards could be used as a model).
- ✓ Penalties for damage or bonus for good performance (example could be a penalty per damaged crop tree or crop tree cut that is not marked. Bonus could be for exceeding certain standards, such as less than 5% crop tree damage).
- ✓ Terms under which the contract is considered complete and performance bonds can be returned.
- ✓ Portion of payment to be reserved until completion (1/3?).
- ✓ Full disclosure of other parties involved in contract (such as a subcontractor to build roads).
- ✓ Liability for contractor to make sure all laws and regulations are followed.
- ✓ Right of landowner to terminate operation for good cause with adequate notice to logger--but pay for what is done.

OPTIONS FOR PROTECTING LAND

Do you wish to continue owning the land?

Yes

- Conservation easement
- Mutual covenant
- Deed restriction
- Lease
- Donation by will
- Donation of a remainder of interest
- Donation of undivided interest
- Management agreement
- Register of critical areas

No

Is compensation necessary?

Yes

- Donation to establish a life income
- Fair market value sale
- Bargain sale
- Installment sale

No

- Outright donation
- Donation by will
- Donation by a remainder interest
- Donation of an undivided interest

Do you wish to restrict future uses when you transfer title?

Yes

- Conservation easement
- Deed restriction

No

- Donation
- Sale

A combination of options may be used to meet your goals for your land.

Adapted from *Conservation Options: A Guide for Maine Landowners*, by F. Mariana Schauffler, edited by Caroline Pryor. Maine Coast Heritage Trust. 1994. The booklet has discussions on each one of the options. For more details about all these options, contact the Maine Coast Heritage Trust, 167 Park Row, Brunswick, ME 04011

Motts' Side of the Mountain: How one family is protecting old growth and viewsheds on a small mountain in Lakeville, Maine[1]

The following profile illustrates how a family can use silvicultural and legal tools to do landscape planning for long-term ecologically-based management and protection of special areas. The Motts are protecting a core of old-growth, buffering it with low-impact management, and assuring that key areas of forest will not get converted by development. Gordon Mott is a forester with decades of work with the Canadian Forest Service (where he studied and modeled the dynamics of spruce budworm outbreaks), the U.S. Forest Service, the Passamaquoddy Tribe, and private practice. He was also a teacher at Unity college and a consultant with the Natural Resources Council of Maine.

At barely over a thousand feet in height, Almanac Mountain doesn't seem to be much of a mountain. From the firetower on top, however, one can see from Eastport to Katahdin, with an impressive view of the numerous connected lakes that include Sysladobsis, Junior, Bottle, and Duck. To the south, below the tower, is a cliff called "the ledges," which also has an imposing view of lakes, mountains, and valleys. And below this is a beautiful mixedwood forest, full of huge glacial boulders.

In some of this forest, there are piles of hemlock bark indicating cutting to supply the leather tanning industry—which had left the area at the turn of the century. On 50 acres of these woods, however, there is no evidence of any cutting. Some of the trees are quite large. I measured a red spruce at 27 inches in diameter, a white ash at 28 inches and a yellow birch at 33 inches in diameter. The site is registered with the Natural Area Program as old growth.

I visited the mountain, with its views and old growth, at the invitation of Gordon Mott, whose family had recently purchased much of this land. Gordon had told me about the old growth for a few years, but it was not until 1998 that I finally had a chance to get a tour of the land. Besides walking though the old growth, I also wanted to learn more about Gordon's plans to balance protection of the public values (connected with the old growth and viewsheds) with the private values of home, farm, and woodlot.

I have known Gordon since the late 1970s. His mention of protecting old growth and managing other parts of his forest with "long-rotation, high canopy closure," made me realize that he had something of importance to contribute to the long-term approaches of low-impact forestry. I had to see it. I took some friends as well.

Finding the homestead

The Mott home seems isolated. It is at the end of a long steep driveway, with no neighbors in sight. Gordon and his wife Ginny (who was an elementary school teacher at the time of the tour) are used to living in isolated places. Before finding their current property, they lived for two years on an island off the coast of Maine. There they pilot tested approaches to enhanced use of island natural resources—including portable sawmilling of island timbers, pioneering steelhead aquaculture in mid-Maine coastal waters, and testing out agricultural crops and techniques for the Island Institute.

[1] First published in the *Northern Forest Forum,* Vo. 7, No. 5, 1999

The isolation of the island suited them fine as it allowed them to raise their sheep and coyotes with minimum hassle. Their three coyotes were born in captivity and placed with them by the National Wildlife Service. The last died a few years ago, aged 17. After living with coyotes, Gordon and Ginny became advocates against coyote bounties. Gordon's e-mail name is "coyote."

After their stint on the island, Gordon and Ginny looked for an attractive place to settle in a climate similar to Ginny's home in Dover-Foxcroft, and somewhat accessible to the ocean and Gordon's roots in New Brunswick. Gordon wanted a mixture of softwoods and hardwoods to manage. They found what they were looking for in 1989 on 106 acres on the south side of Almanac Mountain. This land was formerly part of a 6,800 acre parcel bought from the Penobscot Tribe and liquidated and subdivided by a large developer. Fortunately, not all of Mott's purchase had been stripped, though other lots in the subdivision had been. Indeed, the cutting had begun, and Mott had it stopped upon his purchase.

The Motts did not build here immediately. Gordon still had unfinished business with the Passamaquoddy Tribe, and so they moved their sheep and coyotes to a farm at the end of a dirt road nearby in Topsfield. Gordon did visit and explore his own as well as the surrounding land that went up the mountain. On one such excursion, he came upon a stand of impressively large trees on a steep, boulder-strewn slope. He suspected this was old growth and contacted the owner of the land. The owner acknowledged there was some large timber on the land (loggers had approached him), but the land was not for sale.

Buying the rest

The Motts moved into their new owner-built home on the mountain in 1995. Three years later, Gordon was looking on the Internet to see if another lot was for sale when he found the upper mountain lot on the market. It was a 160-acre quarter section. Twelve acres on the 1047-foot summit were owned by the Maine Forest Service for its fire tower. Three acres were owned by Maine Public Broadcasting for a TV antenna relay. There were also a few lots created to settle wills. This left 135 acres surrounding the summit—including the old growth and "the ledges."

Mott called up the real estate company and found out that an agent was nearby. He jumped in his pickup—"Hell, I chased him"—and found out the land, indeed, was for sale. "I said I was 'interested.'" The owner accepted an offer at the asking price contingent on 60 days to arrange financing. Three of Mott's sons were immediately willing to participate in acquiring ownership. Together, they were able to put up 25% of the purchase price. But, because of federal rules established after the Savings and Loan crisis, the banks required 35% cash. Now what?

"It looked like this was more land than we would be able to acquire," Mott said. "We were acting in our private interest, but were also motivated by our perception of the public interest. This little mountain should not be entirely in private hands. It is the most significant element of the viewscape from a number of lakes. It should not be stripped by feller bunchers going up the hill, as so many other mountains in the area have been. The old growth should not be converted to veneer. It should be kept pristine. Generations of people from a large region here go up to the ledges for the view—especially in foliage season. There have even been marriages up there. This is land that should be in public ownership."

Just as it was becoming clear the Mott family couldn't afford to buy the land, by serendipity, a "green angel" appeared. Mott was talking to a client, "a man of some means with environmental values," who asked what was happening and if he could be of help. He could indeed. Mott worked out a deal. He went through his client, instead of the bank.

"He is an investor who deserves a return on his money," Mott said. The Motts are paying him at competitive rates. He also has security on the investment—if the Motts are unable to pay, he gets the land—an ordinary mortgage arrangement. "More people with money to invest would probably readily engage in such 'green investments' if they were offered competitive returns," suggests Mott, implying that there is an opportunity for interested institutions to facilitate such transactions in an open investment market.

Zoning for private and public benefits

Having purchased the land, the Motts now had the challenge of managing it to ensure that public benefits would not be lost. Under the Mott family's management, all of the old growth is off bounds to any cutting or development. A buffer of non-old growth can be managed, but biological considerations are primary in this zone. What Mott calls "heritage trees" (large, old trees that were formed primarily by natural forest influences) will not be cut—and there will be no-cut zones (around ½ acre) around these trees. Cutting elsewhere in the buffer would "anticipate mortality and be based on biological, not economic, rotations." Mott's management system in this zone, "long-rotation, high closed-canopy forestry" is equivalent to the goals espoused by low-impact forestry.

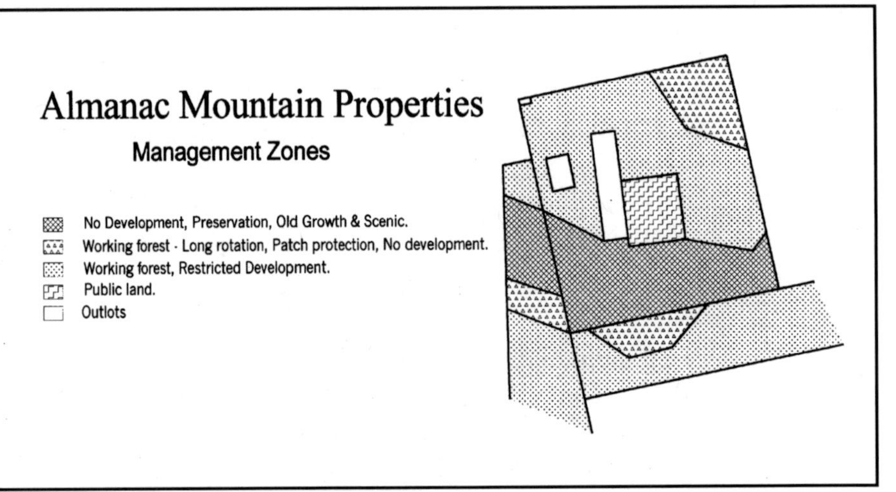

Outside of the old-growth buffer is a managed forest zone, where timber and economics have more importance, but ecological aspects are still considered. Logging will be sensitive to the viewscape and not leave large openings. All trees with cavities are retained and effort is made to restore tolerant species in the regeneration. In this zone, Mott plans to follow "sound, science-based silvicultural standards" and retain all high-quality stems in the growing stock. Here too, minimizing impacts to the residual stand is key to the long-term strategy.

While Mott sees potential for good economic returns in the long term in the managed forest, in the short term they will be cutting small volumes of mostly low value wood. He is up against poor economies of scale combined with low prices. "It's darn difficult to do the right thing to a degraded forest," Mott said. "You're lucky to break even at the front end."

Mott has started logging, utilizing a good local logger with a small bulldozer and winch. The products include firewood for the local market, birch boltwood, hardwood pulp, and several thousand feet of fir and hemlock beam timbers. Mott processes the beams himself on a small band sawmill.

In the forest production zones, as with the old growth and buffers, there will be no development. The Mott family will donate the easement to this right. His neighbors, who own the prominent east face of the mountain, have also indicated a willingness to sell an easement. There will, however, be some limited areas where future development is allowable. Indeed, the family is planning to build a camp on one of these areas. Any structures that they build, however, will not be highly visible from the lakes, and they will screen outdoor lighting from external view as well.

Why easements?

I asked Gordon why the family is trying to sell conservation easements on land with public values, rather than sell all the rights for full fee. He responded that "This is part of our backyard. Part of the conservation easement would be within three hundred feet of our house. What we are seeking is a private/public balance. With an easement, we have standing to enforce agreements. It gives rights to my family (or whoever else might own the land) for the long term to have a say over what the public does right next to them. At the same time, it gives the public a say over what we do with our part of the easement. This leaves checks and balances and creates a buffer for the area of public interest. It's the best place to be. It's win-win."

Mott has concerns that public access to more sensitive areas in the old-growth be restricted until the flora and fauna are cataloged and that public access be on planned routes to avoid erosion and disturbance of sensitive areas. He would particularly like to ensure in the easement agreement that nesting and denning creatures not be disturbed in certain high-risk seasons. The cliffs and boulders create many potential denning sites. He also wants to ensure that hunting and trapping in the natural area be prohibited, while retaining the privilege of hunting elsewhere on the property.

The Mott family has set up a committee to make sure that the intentions of their management plans are carried out for the long term. With a school teacher, builder, surveyor/civil engineer, electrical engineer, and a forester, the Motts contribute plenty of diversity to this committee.

While they are hoping to sell easements to the state through the Land for Maine's Future Fund—the existing public ownership of the fire tower parcel is a plus in this regard—they are open to other possibilities if this arrangement falls through. Land trusts, which could hold easements, for example, could protect the same values. But there are no land trusts in this area. They could, however, help to establish a regional land trust as a repository for the conservation easement rights. Another option, though less binding unless a second party owns and can enforce the terms, is placing restrictions on the deeds.

Mott finds an irony in the opposition of "property rights" groups to public funding of land purchases. "First they said if you want it, buy it. Now the public wants to buy it and they are opposed. It's perverse," he said. "I'm a firm believer in private property rights. 'Property rights' people betray their own principles when they oppose any private property owner's absolute right to protect public values for the future on a willing buyer, willing seller basis."

On a broader perspective, Mott offered: "Like most of what I've been involved with, by luck or something else, this situation found me. I thought I was at the end of my little contribution to how things are done in the woods when I got blessed by the most complicated management proposition I've looked at—in my own backyard. Now, we've got the obligation to set a good example out in public."

*****

13. Options for Small Landowner Forestry Cooperative Organizations for Eastern Maine

Paper prepared by Jim Fisher, Hancock County Planning Commission

Woodlot owner cooperatives are not a new approach in North America. Some regions, such as the Maritimes of Canada, have a long and rich history of woodlot owner cooperation in marketing and managing. In the United States, a new breed of coops is arising that has an interest in marketing and in promoting ecologically-based forestry as well. Some of these coops are getting their members certified as well-managed forests for both recognition and marketing. Some are going farther and setting up value-added opportunities (such as milling and processing of wood) for members. In Sweden, some of the woods coops even own paper mills! There are many advantages to joining ranks with other woodlot owners, but, as this study shows, to get those advantages can require a lot of organizing and hard work.

Purpose

This chapter provides background on the different ways landowners can join together to gain greater power in the marketplace, to add value to wood products harvested from small woodlots, and to achieve greater economies of scale. It was prepared in response to the interest expressed by landowners to a survey conducted by the Hancock County Planning Commission (HCPC) in January of 1999. Sixty-nine respondents (56%) responded positively to the question, "Would you be interested in joining a *Low-Impact Forestry Project* landowner cooperative that included services such as a sawlog concentration yard, kiln services, and marketing assistance?" The percentage rises to 85% when we consider only those persons who provided a response to the question.

In this chapter the term "cooperative organization" is intended to illustrate ways that forest owners can work together--from relatively loose knit networks to organizations that act as a clearinghouse for most or all of a member's purchases and/or sales. This chapter will also highlight the emerging high-value market for wood products that have been produced using certified, sustainable methods. A recent random sample survey of Maine households conducted by HCPC indicates that 85% would be more inclined to purchase a wood product if they knew that it was produced by a Maine woodworker using wood traceable to a Maine "green-certified" and locally-owned wood lot. Approximately one-half of respondents (49.5%) indicated that they would be willing to pay up to 10% more for Maine certified wood products and 10.5% are unsure.

Why Cooperate?

There are both public and private reasons for considering cooperative organizations. From a public perspective, Maine's forestland is undergoing an unprecedented shift in ownership. Recent announcements of land sales suggest that long leases and private ownership by paper and wood products companies may be giving way to smaller land parcels being dedicated to a variety of economic and recreational uses. There are clear public benefits to environmentally sound, yet productive management practices of these parcels. The future of paper and wood products manufacturing in Maine hinges on the willingness of landowners to permit timber harvesting.

From a private perspective, current and future owners of smaller parcels, most of which will ultimately be measured in hundreds or even tens of acres, may no longer have sufficient land to economically manage their property for wood production. Survey data suggest that many small tract owners are willing to consider tree harvesting from their properties provided that the harvesting does not hinder other uses of the property and does not cause long term environmental damage (Widman, 1988, Poitras, 1996).

Cooperative organizations have a rich history in the United States. The classic application of cooperative organization is to reduce costs and expand markets for small producers, who, by themselves, would not be able to secure low prices for their capital and operational inputs nor high prices for their products. The fragmentation of forestland ownership in Maine appears to be conducive to formation of a cooperative mode of organization. However, the benefits to the landowners must be tangible and the uncertainty about negative impacts from cutting on their property minimal. In many cases the effort put forth by the owners must be relatively small as well.

What do cooperative organizations provide?

Maine cooperatives serve a wide range of purposes. Some of the better known Maine Coops include the Eastern Maine Electric Coop supplying electricity to several Downeast communities, the North East Food Coop supplying dozens of Maine community-based food buying groups, and the Potato Growers Coop, which is working to improve quality and markets for Maine grown potatoes. The number of networks, associations and loose knit cooperative organizations in Maine is far higher. Each cooperative organization operates under unique circumstances and consequently has its own character. Even so, the services provided by most cooperative programs fall into one of several categories:

Organizing and Educating. Most efforts begin with organizing participants, and many efforts fail at this early stage. Organizing requires time--very often in the form of evening meetings, phones, faxes and email.

In some instances, relying on existing and well-established organizations can reduce the up front and overhead costs of getting started. For example, a forest network might evolve out of a customer base for a local saw mill. The mill owner benefits by better managing the demands on his facility and the forest owners benefit through a stronger relationship with the mill owner, reduced capital costs, and bartered labor.

Better quality of services may also flow to organized groups. Extension agents, product vendors, lenders, legislators and others can communicate far more efficiently with a group and are far more likely to spend time to explain what they have to offer. As such, organizing and educating a membership has many indirect benefits in addition to preparing for production and sale of wood and wood products.

Capitalization. Most businesses expend tremendous effort in acquiring adequate capital to get up and running. Financing must be directed toward long and medium term assets, such as storage buildings, wood kilns, and large vehicles. Finance for "operations" is equally important and often very difficult to plan and obtain.

Cooperative organizations have been employed throughout the world to ease the difficulty of meeting long-term and immediate needs. In some instances lenders are unwilling to finance small operations with little track record or collateral. If money is lent instead to the cooperative organization with members backing the loan, risk can be reduced to the lender and lower interest rates obtained. Loan guarantee programs from the USDA and others can further reduce interest rates for members.

Purchasing. Some forestry inputs, particularly consumables such as fuel and bar oil, can be purchased in bulk quantities for lower prices. Legal, accounting, management and records keeping services may also be acquired at substantial savings by a cooperative organization. The benefits of participating in purchasing are proportional to the quantity purchased, providing an incentive for members to buy through the organization. The most widespread systems of cooperative purchasing in the United States are probably food cooperatives in their many forms.

Processing. Tools for forest management and wood processing are generally very expensive and may see only short periods of intense use. Small producers generally cannot justify the expense of harvesting, hauling and transportation equipment. Landowners having less than a 10-cord minimum find it difficult to arrange moving their product to market.

Further, wood products have tremendous potential for adding value through post-harvest processing. Practices such as sawing, shaping, drying and finishing are unavailable or uneconomic to small operations, again due to the large capital investments required. Energy costs associated with these tools are also high and often force users into a "demand ratchet", where electric bills remain high even when the equipment is not in use. As such, it is imperative that equipment be operated as continuously as possible. Joint purchasing or use of capital equipment can spread costs sufficiently to justify the investment.

The potential for equipment sharing in forestry is much greater than in agriculture where harvest seasons are compressed and competition for timely use of equipment is high. There are a variety of economic and control mechanisms for sharing equipment, including rationing, renting, bidding and negotiation.

Marketing--broadening market, negotiating higher prices, improving quality, product identification and certification. Small producers often lack time or experience to manage marketing their goods. The functions of a marketing organization include:

- conducting market research to understand customer needs,
- using advertising and other communications to broaden markets into new locations or new customer segments;
- identifying opportunities to earn higher margins for output;
- developing effective brands, certifying quality, and certifying environmental or fair labor practices. More will be said about marketing shortly.

Commitment, Risk and Reward. Finding a good balance between commitment to the cooperative organization and freedom to act independently is key to exploring these operational areas. Suppliers of inputs and purchasers of wood products value predictability of trade as much as the quantity. A cooperative organization that cannot easily predict the amount of purchasing or selling that will pass through the organization in a given period of time due to very flexible membership policies may not receive significant price advantages.

Cooperative organizations that guarantee a volume of trade over time reduce risk for their trading partners and secure better prices--but only by assuming the risk internally. Commitment among the members to supply a specific quantity of wood to the cooperative organization shifts the risk to the members. There are some opportunities for the cooperative organization to mitigate risk by pooling member purchases and harvests and setting production and price ranges. Further, the cooperative organization may be in a better position to secure insurance against trading shortfalls. When an organization is larger and supply chains are distributed, shortages and excess tend to "balance out" more often than in the case of an individual producer.

Three Models of Cooperative Organization

We will now consider three models for cooperation as they apply to forest owners. They will be presented in sequence from relatively loose-knit landowner networks, to Forest Banks and then to active marketing cooperatives. This discussion will conclude with a summary table illustrating some of the tradeoffs between time commitment, resource commitment, risk and profit potential.

1. *Networks*

Changes are occurring on several fronts, including demographics of forest ownership, environmental law and technology, and the interactions of these movements have created new opportunities for organizing productive alliances. Consider the following:

- Forestlands are more fragmented than ever, with trends towards further reductions in average plot size and multiple uses of land. Property changes hands more frequently and owners are less likely to invest in long term productivity.
- Environmental and land use laws are increasingly stringent. The laws are designed to discourage clear cutting, reduce erosion into watersheds, and discourage loss of endangered species habitat and wetlands.
- The science of land use management has leaped ahead with technological advances in remote sensing, global positioning (GPS), database management, and geographic information systems (GIS).
- Information technology has leaped ahead with the Internet, electronic commerce, cellular phones, and faxes.

The organization of work is changing in conjunction with these trends:

- The use of subcontracting, as opposed to direct hiring of employees, is growing rapidly.
- Work at home arrangements are gaining popularity.
- Massive consolidation of some industries is creating niche opportunities for new businesses.
- Globalization of trade has dramatically altered the relative value of skilled and unskilled labor and products.
- Perhaps foremost among the changes in the way we work is the rise of networking.

The network-based cooperative organization is a recent adaptation to these environmental, demographic and technological trends. Consider the impact that these changes can have on forest management:

- By employing GPS and GIS technologies, small forest owners can precisely map their lands with respect to species of tree, slope, water bodies, wetlands, and points of access. Much of this information already exists and is freely available.
- Market prices for wood products can be established through Internet-based bidding.
- Network technologies enhance sharing of harvesting plans along with estimates of yield across many properties.
- Harvesting schedules can be coordinated to maximize use of costly equipment.

- Cellular phones permit owners, workers and mill operators to maintain communications in the field.
- Information can be maintained on a web page indicating reliable suppliers, buyers, partners, prices and services providers.

The heart of the organization is web-like communication. By this we mean that there is no top or bottom--no formal organizational chart designating leadership and roles and responsibilities. There is only a web of connections between individuals interested in striking deals that are mutually beneficial. In the past, these networks tended to be geographically limited. The complexity and investment of time for deal-making prevented small forest owners from getting involved. Now individuals can identify sellers or buyers in minutes anywhere in the country or the world, compare prices, and arrange shipping in minutes.

The network-based cooperative organization does not need to be a technical *tour-de-force*, and may begin with more traditional systems, later evolving as conditions permit. Printed newsletters with helpful information, phone trees, fax broadcasts and regular meetings can facilitate building the network. As participants become more active and more committed, the process of enhancing information sharing, equipment sharing and the like can be introduced.

A coalition of organizations in New York has created a program called Master Forest Owner/COVERTS to network forest owners. So far, 130 "Master Forest Owners" have received training so that they can reach out to their neighbors with ideas for forest management, ecology, soils, communications techniques and more. Trainees benefit from learning new ways to manage timberland. In exchange, trainees bring local credibility to land stewardship programs and can provide high quality outreach at little to no cost to the organization.

Case study – Vermont Family Forests

Vermont Family Forests (VFF) is a non-profit forestry outreach and education initiative whose mission is to "promote the careful cultivation of local family forests for economic and social benefits while protecting the ecological integrity of the forest community as a whole" (Brynn, 1998). VFF is attempting to build local value-added networks to link landowners, loggers, small sawmills, and area craftspeople. The success of the initiative is largely due to Addison County Forester David Brynn and his ability to forge links with local environmental groups and to educate member landowners on sustainable, low-impact forestry practices.

VFF has sponsored numerous workshops on topics as varied as portable sawmill operation, timber grading, riparian zone restoration, and water quality protection. The sessions have been designed to provide the "non-professional" logger with the necessary tools to practice sustainable forest management. VFF encourages landowners to follow their adopted voluntary timber management practices that have been designed to enhance forest productivity and to protect water quality and biological diversity.

VFF has achieved many of the objectives common to a network organization, having connected landowners, promoted education and quality standards among members, and identified value-added markets. Phase one of the VFF program was to develop an affordable model for member landowners to receive independent "green" certification status through SmartWood--a goal they successfully completed in the summer of 1998. In phase two of the Vermont program, "a non-profit business plan will be developed and the potential for construction of energy efficient, wood drying kilns to dry lumber locally will be examined" (Brynn, 1998). The move toward acquisition of a kiln or other "hard goods" is a step beyond the networking model and toward a more integrated cooperative structure. Having laid the groundwork through networking, VFF is poised to go the next step.

2. ***Forest Banks*** Land banks are an evolving response to the shift of land from active production to passive production or preservation. There is a growing desire among landowners to preserve their property, and state policies are being developed that permit reductions in property taxes for land that is placed under restrictive easements.

For example, transfer of development rights (TDR) programs have been employed by community planners to permit owners to recover some or all of the rent that they lose in putting land under protection by permitting a higher intensity of development on other land. A TDR program allows owners to place land in a conservation easement, reduce the taxes on the land and/or gain permission to develop other property at a higher than normal density. This right to higher density development can even be sold to others within the TDR jurisdiction.

The concept of a forest bank is similar. Forest owners place the timber on their land under permanent easement and "deposit" their timber into a bank. The timberland is assessed and a forest management plan is prepared which specifies a harvesting plan. The owner is paid an annual distribution that averages the future stream of revenues projected for the timber. Owners who need revenues sooner can receive a lump sum payment in lieu of the future stream of revenues. As with early withdrawal from a certificate of deposit, the total amount received would be less. It may also be possible to use less-than-permanent easements, with penalties for subsequently removing timberland.

Forest banks enjoy some of the advantages of cooperative organization that were spelled out earlier. If a bank is large, then the flow of products from the bank should command higher prices from buyers. Similarly, contractors who have predictable long-term work plans can manage their costs better and pass savings on to the landowner. This concept is relatively new and requires very careful planning. Questions about who makes management decisions, how wood is graded, and how to incorporate biodiversity planning are among the many that need to be worked out.

Case Study –CCED Clinch Valley Project, Virginia
The Nature Conservancy, now almost 50 years old, has traditionally worked to preserve ecosystems through land protection, conservation easements and ownership. In 1995, recognizing the need to work with more private forest owners and understand sustainable wood harvesting, they created the Center for Compatible Economic Development (CCED). The Center is an incubator for profitable ventures that improve forest quality and protect ecosystems. Their first partnership is in the Clinch Valley in central Appalachia. CCED has formed a "Forest Bank", in which the customers deposit their timber and make "an ironclad promise" that the land will remain forested forever. Future harvests will be selective and geared toward overall health of the land. The Clinch Valley project targets relatively small, private, non-industrial owners who wish to maintain healthy forests. While the project is still too new to assess, it will be worth tracking in the future. The Nature Conservancy is already investigating other locations for forest banking, including Indiana and Michigan.

By depositing timberland in a bank, forest owners maintain a flow of income from their timber while maintaining the quality of the landscape for residential, recreational or environmental purposes. In return, owners give up their right to clearcut timber, subdivide the property for non-forest uses and so on--or at least would be required to pay penalties for these changes. As such, this form of cooperative organization is fairly passive with respect to the owners, who may not be required to devote much time to the organization. This type of organization still requires a strong commitment, with owners relinquishing some options for subsequent use of their land.

Some issues that must be answered in forming a land bank include:

- managing differences in amount and quality of land owned by depositors;
- meeting short-term revenue expectations during the initial period of recovery for young or poorly managed forests;
- managing different levels of commitment to sustainable practices among depositors; and
- negotiating with wood cutters, mills and end users.

3. *Marketing Cooperatives*

Marketing cooperatives have a long history within the United States and are employed broadly throughout the world. Even so, marketing itself and its place in cooperative organizations is poorly understood. The marketing function for a forest cooperative organization can be very limited, such as identifying the best price for logs. On the other hand, the organization may be involved in a full range of services, including research, product development, branding, advertising, sales-operations, business management and customer relations.

In order to establish some breadth of possibilities, a few basics are in order. Marketing professionals often resort to "four P's" and "three C's" to summarize critical issues in turning ideas into valued products or services.

Four P's:

1) *P*roduct – a description of the goods or services that you want to sell. Questions to ask include how your product is different from that of your competition, whether you can increase profits through additional materials processing, and how you handle changes in consumer taste.
2) *P*rice – a projection of the likely market prices as well as models which indicate the "elasticity of demand" for a given product, or the impact of increasing price on the amount that will be sold.
3) *P*lacement – The geographic regions, intermediate traders, end users or other projected market segments that you anticipate will purchase your products or services.
4) *P*romotion – The means by which you will notify customers of your products or services. Promotion includes paid advertising, trade shows, door-to-door sales, press releases, product samples, discounts, coupons, and a host of methods for getting the word out.

Three C's – Critical threats to a successful marketing plan:

1) *C*ustomers – The business or individuals whom you think will buy your products. In the case of pulpwood, a relatively small number of customers are very powerful in setting prices and delivery schedules. The situation for other wood products is far more complex. Difficulty in identifying and building relationships with customers is a very common problem for small businesses.
2) *C*osts – A marketing plan must anticipate costs of production and overhead and indicate how these costs will be paid. The costs may involve cash, but often as not they are in-kind contributions of time and materials. Cooperative organizations often have managerial costs that members are reluctant to pay. Many volunteer run organizations extract heavy unpaid commitments of time from a few members who eventually burn out.

3) *C*ompetition – Competition includes existing producers as well as future entrants into the market. Competition for products that are undifferentiated commodities is based almost entirely on prices. This is not a desirable position for the sellers unless supplies are extremely constrained or production costs are very low relative to your competitors. Pulpwood is very close to being a commodity, though there are some quality differences among grades of wood.

Recently there have been some indications that consumers are willing to pay a premium even for commodities if they are assured that the methods of production are less harmful to the environment or society. Indications of this trend include the movement to boycott products from companies that use child labor, and the premium prices paid for electricity that is generated from sustainable or low impact sources. Wood products--including specialty lumber and shaped or finished pieces--are more often differentiable. They can earn better prices if quality is superior or production methods, which are thought to be more acceptable, can be certified by a well-respected third party.

Cooperative organizations that focus on marketing should assist members in preparing business plans that address the four P's and three C's, and then in implementing these plans. Unlike the network organization, the marketing cooperative organization is likely to have a formal structure and related expenses for contracted services. Members must make some level of commitment of their resources to pay overhead expenses. Unlike the Forest Bank, where members may commit very little time to operating the bank, the marketing cooperative requires time and effort from its members.

Case Study – Forest Products Marketing and Management Association/Cooperative, Inc.

The Forest Products Marketing and Management Association/Cooperative, Inc, once based in the Dover-Foxcroft area, provides a wealth of insight into the formative process for a marketing cooperative as well as the challenges of running a cooperative business during difficult economic times. The tax-exempt association came together first, and membership grew steadily to a peak of 150. Members participated in educational programs, networking, and advocacy for forest owners. The association won a 3-year, $70,000 grant to support start-up operating costs and the purchase of a small amount of equipment. The grant allowed them to write a business plan that received recognition but fell victim to high interest rates and low wood prices in the early 1980s.

The Association in turn formed the basis for a cooperative that started with 15 members and grew to a peak of 85 members representing 12,000 acres of forestland. Even at its peak, most of the members of the cooperative lived within a 40-mile radius. The cooperative's goals included:

- increasing market power to improve their ability to negotiate input and product prices;
- pooling resources for harvesting, processing and transporting wood; and
- identifying new value-added products.

The Association/Coop managers initially lived on grant support, then sold forest management services during latter years. Members could borrow money, but only one loan was taken. The managers assisted growers to sell their wood, but the level of transactions was not adequate to support staff in the absence of subsidies. A number of hard lessons were learned in the process:

- The Cooperative couldn't beat the market prices, and in some instances buyers discriminated against cooperative members in setting prices. The coop even faced threats of law suits from large landowners.

◆The structure of the Association and Cooperative was loose, with $25/year dues and freedom to play cooperative prices off against the competition. Many members were not sufficiently committed.

◆The grant consumed a lot of time and moved the organization toward top-down communication. The experience of organizations being undermined by well meaning, external support is actually quite common.

◆Due to small size and low level of commitment, the cooperative couldn't deliver the quantity of wood promised to a buyer. Being tagged an unreliable supplier forced prices still lower.

◆Transportation was a big problem. The typical lot was too small to fill even the minimum truck size, requiring one truck to take a route through a number of small wood lots.

◆Many cooperative organizations fail to specify an "out" clause in the event that they become insolvent. The managers have found it difficult to disburse remaining funds.

◆Despite their collapse, organizers of the Association and the Cooperative see hope in the concept and suggest that this may be a better time to start. Interest rates in the early 1980s ran as high as 20%, nearly three times current levels. Wood prices are significantly higher, particularly wood that has been cut and dried. There is now measurable consumer interest in certified forest products, leading to the potential of better profit margins. Large lumber retailers like Home Depot are introducing new lines of certified products. Pulp mills are buying more on the open market than ever before. The circumstances of many small forest owners may have shifted as well, with increasing interest in long term management and less dependence on immediate profits.

The Forest Products Marketing and Management Association/Cooperative, Inc., did a great deal of experimentation and had some modest success. Despite their ultimate collapse, their model of starting with an association that acted as an umbrella for a cooperative has merit.

Case Study - Sustainable Woods Cooperative

A group of 120 southwestern Wisconsin landowners have joined together to manage, harvest, process, and market value added timber crops in what director/forester Jim Birkemeier phrases a "new generation" cooperative. To date (1999) the Sustainable Woods Cooperative (SWC) members have brought somewhere in the vicinity of 30,000 acres to the cooperative landbase. Charter members pay a $100 fee to join the cooperative with the understanding that their membership promises complete sustainable management of their woodlot: from initial woodlot assessment to timber harvest to value added finished products. The primary objective of the SWC is to identify the most profitable markets for certified wood that can be produced from the mature trees that have been harvested by trained loggers from members' woodlots.

Adding value to their already valuable standing timber is what the SWC will target as they strive to increase economic returns from current harvests and to manage member owners' woodlands for the future. The SWC asserts that they can obtain more money from a smaller concentration of select trees as compared to selling larger quantities of raw sawlogs directly to a sawmill. They maximize profit from each high-grade tree via custom sawing sawlogs using a small Woodmizer bandmill, drying the wood in a solar kiln, and producing high-value finished products in the coopertive's woodshop, such as flooring or cabinetry stock. The Wisconsin cooperative is equipped with small-scale machinery, a concentration yard that keeps track of every log it receives using a computer bar code system, and a barn sales office for marketing their finished products. SWC is headquartered on Jim Birkemeier's family-owned 200-acre farm in Spring Green, Wisconsin.

Discussion

Employing cooperative organizations for reducing production costs and increasing sales volume or price is a promising avenue for small forest owners. Many find that the status quo offers them few viable opportunities for managing their land and earning a significant income. The combination of not having current or reliable information on markets, paying higher input costs, and receiving lower prices puts the small landowner at a tremendous disadvantage. Forming a cooperative organization is one means for overcoming these disadvantages and exploring new ways to improve and profit from the forests.

The three models of cooperation presented in this overview are by no means the only alternatives. They are provided here to highlight some of the dimensions that are possible given limited time and resources. Table 1 compares the three organization types across a variety of issues discussed in this paper.

Table 1 Comparison of Cooperative Organization Options

	Network Organization	**Forest Bank**	**Marketing Cooperative**
Time Commitment	Medium	Low	High
Resource Commitment	Low	High	Medium
Organizational Structure	Informal, Inclusive	Structured, Exclusive	Democratic, Mixed
Organizing and Educating	High	Medium	High
Capitalization	Low	High	Medium
Purchasing	Medium	Low	Medium
Processing	Low	Low	Medium
Marketing	Medium	High	High
Risk	Low	Medium	High
Reward	Medium	Medium	High

These case studies illustrate the potential for success and also the potential for calamity. Success in small-scale cooperative organization is often subtle--so much so that members may not recognize the contributions that the cooperative organization has made. This is particularly the case when there are no clear tools for measuring impacts, such as for educational and organizational programs. The contribution of services that directly impact the bottom line, such as discounts on purchase of consumables and increments to prices of goods sold, are tangible and are more likely to sustain participation.

Some common pitfalls of the cooperative structure include loss of focus, shortage of volunteer time, lack of resources, inadequate financial commitment, lack of skilled leadership, difficulty negotiating with large buyers and susceptibility to market down-turns.

The most significant pitfalls may develop from good intentions. The Forest Products Marketing and Management Association Cooperative received a grant which encouraged them to expand their hired staff and to explore new lines of business. The impact of receiving so significant a grant was to change the focus and the structure of the cooperative. The structural changes in particular may have undermined the cooperative's long term success.

Getting Started. The case studies suggest that some cooperative organizations, such as the Forest Products Marketing and Management Cooperative, begin with unmet needs of landowners and a conviction that untapped market opportunities exist for forest products. Others, like the Clinch Valley Project emerge from opportunities created by changes in environmental and land-use law or changes in consumer preferences. These projects may be led by people who are not themselves forest owners. Each of the groups emerged through extensive discussions, organizing efforts by community leaders, and hands-on research.

Research. Research is an important step to: building a cooperative organization, identifying why the organization is needed, and gaining support from members. Public opinion surveys, literature reviews, and market research are helpful, but may not garner support from independent landowners. Cooperative organizations hold much of the information that they need to organize and launch programs, but the informal and unstructured nature of communications may slow progress and lead to unanticipated conflicts in understanding.

"Participatory action research" is one process for diagnosing environmental conditions and identifying solutions through organizational strengths (Korten, 1984). This hands-on form of research can begin with discussions among landowners about problems that they are confronting--such as difficulty securing good prices for their products, or impacts on property values and property taxes of rural sprawl. Turning these discussions into a slightly more structured research process can yield insight into the underlying issues and build consensus around solutions. The meeting to discuss cooperative options among Downeast landowners is based in part on this method.

Success and Planning. To quote the mayor in the musical, *The Music Man*, "There is nothing so successful as a success." Most organizations need a jump-start, and cooperative organizations are no exception. Immediate successes for the membership may include path breaking research that leads to:

- finding lower prices for inputs,
- recognition and support from state leaders or agencies,
- presentations from experts that help members to improve their work, and
- successful social or fund-raising events.

Winning grants can serve as an immediate victory, but experience suggests that most grants take a considerable amount of time to win, are difficult to sustain, and may distract the organization from its original goals.

In the medium and long run the importance of business planning cannot be over-emphasized. The cooperative organization is well served to discuss and inscribe long-term goals, tangible objectives with deadlines, and strategies for accomplishing those objectives. Without these baseline documents, the cooperative organization is prone to wandering through a variety of schemes without a sense of direction or measure of accomplishment.

References

Brown, Samuel. Personal interview conducted 4/8/99

Brusila, Barbara. *A Procedure for Marketing of Forest Cooperative Members.* Unpublished Masters Thesis, December, 1983 UMO Fogler Library and UMO Graduate Student Library

Brynn, David. Vermont Family Forests: *Green Certification Project.* 1998 Addison County Forester, 1590 Route 7 South, Middlebury, VT 05753

Dirkman, John, *A study of Maine's cooperative forest management program* [Augusta, Me.] Management Division, Maine Forest Service, 1983 UMO LEG Stacks SD413.M2 D5 1983 12 p. charts ; 28 cm

Korten, David. *People-centered development: contributions toward theory and planning framework.* West Hartford, CT: Kumarian Press, 1984 UMO HD75.6 .P46

Locke, Ronald. Personal interview conducted 4/8/99

Master Forest Owner/COVERTS Information listed on the internet at http://www.dnr.cornell.edu/ext/mfo/

Poitras, Ron. *Results of Survey of Small Woodlot Owners in Hancock County.* Hancock County Planning Commission Library. 1997

Widman, Richard H. and Thomas W. Birch. Forest-land owners of Vermont – 1983. Resource Bulletin NE-102. Broomwall, PA: USDA Forest Service, Northeastern Forest Experiment Station. 89 p.

*****

10 Guiding Principles for the Downeast Low-Impact Forestry Cooperative

By Geoff Zentz and Ron Poitras of Hancock County Planning Commission

The quotes listed below each of the guiding principles were taken directly from the recorded proceedings of the exploratory meeting on forestry cooperatives held with landowners in Ellsworth, Maine, on May 15, 1999.

1. Member landowners will commit to following the agreed upon standards and practices of low impact forestry for the management of their woodlands.

"This statement represents very good principles, I don't see how anyone can object. The question is how will we get there?"

"It makes you feel better. That is a big aspect of what we are doing."

2. Cooperative will provide continuing education opportunities for landowners, foresters, and loggers on LIF theory and techniques, available markets, and forest ecology.

"We should definitely include landowners in the training as well as foresters and loggers."

"The cooperative could help with advertising and marketing for loggers"

3. Cooperative will process and market members' wood for highest and best use.

"We need to look at existing markets and create new markets."

"Right now most wood is going to market not at its highest value. I am interested in that. It is important to me because I sell a lot of wood and see a lot of wood wasted. I do not like that."

4. Cooperative will research and identify current niche markets and make this information available to member landowners prior to a timber harvest being conducted on their land.

"The Cooperative could keep an available markets list- something that members can access. I think that access to this information when you need it is important. A page or something. You know where the market is and what is going on."

5. Cooperative will monitor ongoing harvest on members' land, including periodic site visits, and soil and stand damage assessment.

"I think perhaps that 'supervise' is the wrong word. Once the job is completed or at a halfway point, someone would go out and check on the quality of the job."

6. Cooperative will be responsible for all wood transportation and processing details, including trucking, pulpwood sales (possibly through a concentration yard for higher volume), tracking certified logs through sawmill, and marketing finished, value-added products. Members will be able to follow the path of their sawlogs from woodlot to final sale.

"I would want to follow one of my logs through the whole process rather than sell it to a mill somewhere that is just going to mill it off as fast as they can."

"I think that the identification systems during the value-adding, isn't that the crucial point of maintaining an identification of these low impact harvested logs as they are produced into lumber to get your increased value."

7. Cooperative will provide members with a list of loggers and foresters who are trained in LIF. The list will include type of equipment used by individual loggers, and rates for loggers and foresters.

"I have had difficulty finding a forester in the first place, and, secondly, to find a forester who knows something about low impact forestry."

"Finding a forester who is into LIF is hard--I don't know who to go to. If this service existed to help out that would be good."

8. Cooperative will assist landowners in planning for long-term forest management. Coop will not pursue markets or timber sales that are motivated by short-term profit at the detriment of sustainable management.

"Rather than say a particular practice is not acceptable-- there should be long-term management plans."

"In a situation where the timber market is very good right now and we choose to deviate from the forestry plan to meet the market, we are not meeting the by-laws of the coop and we are out."

9. Cooperative will pursue direct sales to local markets that strengthen the Downeast economy and communities wherever possible. If a solid, local market cannot be located, wood will be sold for highest and best use elsewhere.

"I like supporting smaller industry. When the person running the store has an interest in it and really cares what they are doing. I would much rather support that kind of business."

"My interest in this cooperative is twofold. I have a woodlot and I am a builder. I prefer to use local, low impact harvested wood. Strengthen the local economy-- the money stays right here. So much better."

10. Cooperative will develop "brand" awareness for LIF wood that is associated with quality, environmentally-conscious, and locally-produced wood products.

"This ...... is exactly why I am here. I would like a better market for the average guy."

*****

14. As if the Future Mattered...

The biggest challenge for making low-impact forestry work is not our level of scientific understanding or technological sophistication. Science and technolgy serve societal goals. The biggest challenge is to have a society with long-term values.

Priorities?

Low-impact forestry is multi-generational forestry. It is forestry as if the future mattered. The primary concern is to protect the biological integrity of the forest because the forest is part of the life-support system upon which our society depends. Economic goals are pursued only insofar as they do not interfere with this primary goal.

Unfortunately, our larger society has different priorities that conflict with the priorities of low-impact forestry. In this larger society, the forest is seen as a resource of the economic system. The primary goal for the management of this resource is to get an acceptable return on investment. Ecological or social goals, in this context, are pursued only insofar as they do not interfere with this primary economic goal.

In pursuit of this primary economic goal, landowners have:

- liquidated forests and subdivided the properties to maximize short-term returns; or
- highgraded the forest and left it understocked (taken the gold from the mine and left future loggers the shaft); or
- simplified forest ecosystems (with clearcutting, herbicides, and plantations) to produce a crop of trees on short rotations. To some extent, the promise of high yields in the future from these "intensive practices" are used to justify high levels of cutting now. This is called the "allowable cut effect," "accelerated cut effect," or ACE.

To calculate high returns on investment for such activities, those doing the calculating have confused income with capital depletion. The above strategies are not biologically sustainable. These strategies do not save the parts or protect the processes that allow whole forests to persist. The key to making low-impact forestry work for the long term, therefore, is in the priorities of those who have responsibility over the forest. These priorities need to be passed on from generation to generation--they must be part of the culture.

Science?

Landowners need good information about the forest ecosystems with which they are working. Unfortunately, until recently, forest scientists have been more focused on fiber production than ecosystem maintenance. What was considered "scientific" management in the past, is sometimes considered unscientific in the present. Scientific understanding of forest stand and landscape dynamics is changing. For example, what was once thought to be good for wildlife ("edge" for game for example) is now considered a problem for other wildlife (fragmentation for interior species).

Wise scientists, such as David Perry (see chapter 3), recognize that forest management is a long-term experiment. Since the consequences of being wrong on a big scale can take centuries to remedy, he recommends following a precautionary principle. "My own feeling," he stated, "is that experimenting with different management approaches is a good thing--it's how we learn. But prudent behavior dictates the need for guidelines; filling landscapes with an experiment that fails would not be good. One rule of thumb I favor is the more a given management approach departs from the natural forest structure, the less area it should occupy (at least until its stability is established, which could take decades or centuries)." Low-impact forestry is predicated on that sense of prudence.

Perry advocates "adaptive management," which has two major principles. "One is that you maintain the options to adapt, and maintaining the options to adapt means, essentially, that you maintain diversity. The second idea is that [...] you have a clear vision of where you're headed and what you need to achieve on the landscape so you can come back and look and see if what's been done is taking you in the direction you want to go." LIF also encourages clear guidelines and monitoring to see that guidelines are met.

Non-prudent science

"Foresters have been consistently proud of and comforted in the assertion that forestry is a scientific enterprise. It is a claim that has been used historically, at least since the days of Bernhard Fernow's experience in the Adirondacks, as a first line of defense in times of conflict and controversy.

Mr. Fernow was the dean of the Forestry School at Cornell University in the first decade of the 20th century. He was demonstrating good forestry around Saranac Lake by clearcutting hardwoods to favor and nurture the commercially superior spruce; he installed a cooperage factory to utilize the low-grade hardwood. The local citizens who appreciated the beauty of the mountains objected vigorously. In their complaints, they used such terms as 'forest devastation.' Today we would call such malcontents environmentalists, I suspect. Mr. Fernow demurred, suggesting that the lay citizenry simply did not understand the science of silviculture; he was privately outraged that his professional, scientific expertise was so heavily subject to discount, indeed override. For overridden it was. The governor of New York, notwithstanding some historical evidence of inebriation at the critical political moment, set Mr. Fernow's subsequent annual budget at zero, oversaw the closure of the Forestry School, and witnessed the dean's departure for Canada."

From "Scarcity, Simplicity, Separatism, Science--and Systems," by R.W. Behan, in *Creating a Forestry for the 21st Century: the Science of Ecosystem Management*, edited by Kathryn Kohm, and Jerry Franklin, Island Press, 1997

Markets?

Even those landowners with an intention to do "good management" are confronted with powerful short-term economic deterrents. Initial cutting to restore highgraded stands requires removing poor-quality trees and retaining high-quality trees. Poor-quality trees, however, usually go to markets that qualify as commodities, such as pulp. Pulp and paper companies in Maine are competing in a global economy where, if they raised the price of their product to reflect increasing costs of wood, they would lose sales to lower-cost competitors. Some of this competition is operating where there are low costs for labor, few environmental restrictions, and major government subsidies.

To compete, therefore, the Maine-based companies try to cut their costs as well. To the degree that the companies are very large, they have leverage over both local government and local economies. Areas where they have been able to cut costs in Maine include taxes, regulations, energy, and (especially) purchased wood. When the mills cut the cost of their purchased wood, however, this means that woodlot owners get less money to compensate for their management and loggers get less money for their labor. With few competing local markets for the wood, woodlot owners and loggers have little choice but to take the price offered.

Reforming the negative impacts of the global economy is surely not an easy task. In the short term, landowners are finding they can improve their returns from smarter marketing--including finding market niches or even adding value through milling before selling. Sorting yards and landowner coops increase the chances of better economic returns so that low-impact forestry can be more viable. If society better valued biodiversity, aesthetics, or recreation--rewarding practices that enhance those values, and penalizing practices that harm those values--low-impact forestry would have major economic advantages.

Labor and machinery?

Logging contractors, in general, are paid by the volume for wood cut. They have, quite logically, invested in technologies more suited for productivity in removals (for which they are compensated) than for the quality of residuals (for which they are rarely compensated). Contractors have to cut a lot of wood to pay for these expensive machines. To pay for the logging and management, therefore, both landowners and loggers often feel compelled to remove too many trees--especially too many potential crop trees that would have been better left to grow. The result, too often, is mediocre, rather than good management.

Logging technology that has evolved for productivity of removals often is inappropriate to low-impact management. Unfortunately, because big machines are what loggers happen to have, the management technique and the forest structure are adapted to fit available technology. Low-impact forestry works best if there are innovative ways to pay labor to reward careful logging. As a demand for lower-impact practices increases, manufacturers should respond with innovative designs to meet those demands. The fact that innovation has not happened on a large scale yet is telling. Many low-impact loggers now are tinkering with older equipment (adding radio-controlled winches, for example) to get it to do the kind of work they need.

Management continuity?

Despite such deterrents, there are still landowners who have improved the stocking and quality of their forests for many decades. The fact that they have done long-term forestry proves that it *can* be done. Unfortunately, there is no guarantee that succeeding landowners will continue those practices.

In the last decade, huge tracts of land in Maine have gone up for sale, some from families or businesses that have owned the land for half a century or more. The buyers are often investor groups that have a time horizon of little more than a decade. The idea of these investors is to cut wood (giving an annual dividend) and sell land (giving a capital gain). Theoretically, these landowners have an incentive to sell their land for more than they paid. This would seem to be an incentive for leaving the land better stocked than when they bought it. Indeed, if property prices better reflected timber values (so that the timber liquidation value was not greater than the property value), a short-term ownership horizon might not necessarily lead to poor forestry. Simply by subdividing the land, however, owners of large holdings can increase their real-estate values without needing to improve their forests. The parts are worth more than the whole.

Low-impact forestry encourages legal and cultural ways to ensure that long-term land management continues after the current owner sells out or passes on. Gordon Mott (see chapter 12) for example, has his land zoned--some for full protection, some for low-impact forestry. This plan must be followed after he is gone, and there is a committee set up to ensure continuity. Mel Ames (see chapter 11) has put his land in trust, with family members to make management decisions when he is gone. There are no detailed management guidelines. To the degree that Mel was successful at instructing his children in the ideas and techniques of his management, his woods will continue to be well tended. For a safety net, he helped pass a local ordinance in his town that prevents forest liquidation.

Landowner organizational structures?

Some landowner structures, such as land trusts, large families, tribes, institutions, and government, have less of a tendency for fast turnover and less need to maximize short-term returns. Indeed, some of these organizations have a mandate to manage for the long-term. The risk in such structures is that those individuals with the responsibility for managing the forest will succumb to temptations to benefit themselves in the short term, despite their mandate.

The 30,000-acre Baxter State Park Scientific Management Area, for example, has a mandate to do exemplary silviculture. Yet, the first few years of management (during the 1980s) of this demonstration forest were marred by sloppy high-grading and commercial clearcuts during a period of bad markets. Forest activists sued to stop the cutting. The result of the suit was that the Park hired a new manager (the land previously was managed by the Maine Forest Service) and set up an oversight committee. Management has markedly improved since then. Though most of the cutting is by mechanized short-wood methods, the manager, Jensen Bissell, is experimenting with smaller low-impact machines and hand cutting as well.

The Menominee Tribe of Wisconsin has a record of forest improvement for nearly a century and a half on their more than 220,000 acres of commercial forest. This is remarkable because the Tribe has had serious poverty, alcoholism, drug addiction, and other problems that would lead to temptations to liquidate and subdivide. Indeed, over their history, there have been periods when they did cut too heavily and they did sell off lots around a lake. There have been times when the federal government looked over their shoulders--or even directed the cutting. Sometimes this federal involvement was a disaster--especially when the government thought that heavy cutting was the equivalent of good forestry.

Despite these episodes, the Menominee have more standing timber, of better quality, now than they did 145 years ago. In the last several decades, the cutting has been under the control of the Tribe. The Tribe's management is informed by science, but it is guided by tradition. They maintain continuous inventory plots and have a site classification system based on understory vegetation to determine the most appropriate tree species for the site. In some cases, they have used clearcutting and herbicides to convert stands to species that, they believe, ought to be dominating. Most of their cutting, however, is by selection.

One sign of the Menominee commitment to sustainable forestry is the relation of the forest division to the sawmill division of the Menominee Tribal Enterprises. The Tribe does not let the mill drive the logging. The mill has to live with what is available from the forests. The Menominee have been able to keep their priorities so far, despite strong pressures to change. The complex nature of tribal politics and a strong sense of tradition are two explanations for this continuity.

The point of these examples is that even organizations with a long-term perspective need long-term vigilance. There need to be built in checks and balances and oversight. Above all, there need to be individuals who have the wisdom and commitment to make sure that the extraction of timber to meet current needs does not endanger the ability of future generations to meet their needs as well. There needs to be a society that produces such individuals.

Sustainable?

Human beings are capable of very long-term management. Some forest areas of Switzerland have been selectively managed for over 600 years. Regions of China have been farmed for over 4,000 years. In regions of New Guinea, tribes have managed both vegetables and forests with slash and burn agriculture for thousands of years. By managing responsibly within their biological income in the present, these cultures have ensured that future generations can do the same.

In the United States, forestry's economic context operates on a much shorter time horizon, with a poorly developed sense of limits. We live in a society which heavily discounts the future. Next quarter's earnings matter more than next century's forests. While policy planners sometimes project trends of demand for 25 or even 50 years (often to justify the "need" for more intensive forestry practices), they never project such trends ahead a thousand years, let alone a few hundred. If they did, the absurdity of projecting geometric growth forever into the future would become evident.

According to US Forest Service inventories, between 1959 and 1995, the rate of cutting in Maine increased by an average of 2.1% per year. In 2995, therefore, to continue this trend for a thousand years, the cut would need to be 1,061,065,239 times as high as in 1995. This translates to 1,893,939 cubic miles of solid wood that must be cut towards the end of the millennium. Since the forest area in Maine is roughly 26,464 square miles, the equivalent of a solid chunk of wood 137 miles high over the entire area of Maine's forest would have to be harvested in the year 2995. Even with genetic engineering and high-potency chemical fertilizers, the challenge of growing trees into outer space might prove too much for future generations. Geometrically increasing the cut is not sustainable. The sky *is* the limit. Indeed, the Maine Forest Service projected the 1995 level of cut *with no increases* 50 years into the future and found *that* level of cut was not sustainable.

In the past, when Maine got overcut, loggers moved on to new frontiers to the west. Forest area and volume in Maine recovered somewhat as farms were abandoned and loggers found bigger and better wood in Minnesota, Wisconsin, Michigan, or Oregon. Now, our wood-supply basket is the entire planet. There are still some frontiers in Siberia or the Amazon, but when we reach limits this time, it will be at a global, not local, regional, or national level. There will be no new frontier of undeveloped forest to conquer on earth. Astronomers have not found evidence of waiting virgin forests on any nearby planets. If we are inevitably going to meet limits of consumption, it would be better to do so when we still have more natural and wild forests than when we have lost them.

Forests can sustain themselves quite well. Sustainable forestry implies a relationship between society and forests. If the society is not sustainable, neither will be the forestry. Since forestry is an experiment, we need unmanaged forests as a control. We also need them as an example for how forests are structured and how they function across the landscape. And we need them to protect all the habitats and all the species that would occur naturally on the landscape, including ones sensitive to human encroachment.

Reforming the forestry system, and the larger economic system in which forestry operates, to favor wild forests and long-term, low-impact forestry, is a daunting task. The first step is to change direction--to act as if the future truly matters. Other actions (such as changes in governmental forest policy) will follow as a result. If we do not change direction, however, we will, as an ancient Chinese sage warned, wind up where we are headed.

Appendix I: LIF Equipment Suppliers

Compiled by Geoff Zentz

Distributor:	Equipment:
NovaJack Quebec Tel: 1-800-567-7318 e-mail: info@novajack.com website: www.novajack.com	**Logging trailers, arches, gas winches, skidding cones.** NovaJack has a wide range of equipment for low-impact forestry for those working on a small scale, including safety equipment.
Payeur Inc. 5379, King Street East Ascot Corner, Sherbrooke (Quebec) JOB 1AO Tel: (819) 821-2915 Fax: (819) 820-0490 e-mail: payeur.distribution@videotron.ca	**Metavic's "The Forester"-** log loaders and trailers with hydraulic winches. Designed for selective harvesting, the "Forester" allows the operator to maneuver easily throughout entire woodlot without the need of many roads.
Silvana Import Trading Inc. 4269, rue Ste. Catherine Ouest, Suite 304 Montreal, Quebec, H3Z 1P7 Tel: (514) 939-3523 Fax: (514) 939-3863 website: silvana@total.net	**NIAB 5-15 Tractor Processor**-small processor that winches, limbs, measures, bucks, and sorts small diameter logs. It is designed to hook directly to a farm tractors 3-point hitch and is powered by the PTO. (Manufactured in Sweden)
Nova International Ltd. RR #2 Windsor, NS BON 2TO Tel: 1-888-404-6682 Fax: (902) 798-8585	**PATU loaders and trailers, PATU stroke delimber for tractors.** Low impact forestry equipment for the farm tractor, PATU's designs are for direct mount to the tractor's 3-point hitch. The stroke delimber can be operated directly from the A-frame of a PATU loader or hook to the 3-point hitch if there is no loader. (Manufactured in Finland)
J.S. Woodhouse Co., Inc. PO Box 1169, 1314 Union St. West Springfield, MA 01090 Tel: (413) 736-5462 Maine tel: (207) 873-3288	**PATU loaders and trailers, PATU chippers, PATU processor/harvester head, FRANSGARD timber winches** (Finland)

Prescott Mouldings and Wood Products 489 Archibald Brook Roa. Middle Musquodoboit Nova Scotia B0N 1X0 Tel: (902) 384-2720 e-mail: av987@chebucto.ns.ca	**forestry equipment for the all-terrain vehicle and other assorted logging equipment**
Marfor Equipment Ltd. 101A Moss Avenue Fredricton, N.B. E3A 2G2 Tel: 1-800-565-6446 Fax: (506) 450-3706 e-mail- marfor@nbnet.nb.ca	**"pied piper bobsleds"** a low to the ground, 4 runner woods-sled that can be hooked behind either an ATV or a small tractor
Union Farm Equipment 1893 Heald Highway PO Box 155 Union, ME 04682 Tel: 1-800-935-7999 e-mail: ufeinc@ime.net	**PATU log loaders and trailers, tractor winches (Norse, Fransgard, Fisher Iron Works), "Stumpster" stump grinders, Tractors**
Thomas Bandsaw Mill PO Box 95 Brooks, ME Tel: (207) 722-3505	**Thomas Log Arch-** the old log arch idea that is equipped with large used car tires and designed to be used behind an ATV, tractor, or horses.
Addington Equipment PO Box 179 East Dover, VT 05341 Tel: (802) 348-7872	**Tractor Grapple, Brush/Grapple Rake-** both implements are designed to hook onto the standard farm tractor. The grapple is designed to fasten directly to the 3-point hitch and is available with a winch option. The grapple gives the operator the ability to either lift logs and move them in the air or skid logs by hooking one end and towing. The brush/grapple rake replaces the bucket on the front end loader and is used to push slash.
Carl W. Neutzel Services, Inc. 2648 Openshaw Rd. White Hall, MD 21161 Tel: (410) 329-6791	**Skidding winches and grapples for all size equipment, forest trailers for trucks and farm tractors, Stroke delimbers, Keto harvesters, Knuckleboom loaders, firewood splitters, processors and conveyors.** "Equipment for responsible forestry practices and resource recovery."

Northeast Implement Corporation PO Box 402 Spencer, NY 14883 Tel: (607) 589-6160	**Valby disk chippers, Farmi logging winches, Forwarder trailers, Knuckleboom loaders.**
The Lewis Winch 221 North Central Avenue, Suite 235 Medford, OR 97501 Tel: (541) 664-8616 Fax: (541) 664-8619	**Chainsaw winches available in three models-** the Lewis Winch offers portable pulling power rated to 4,000 lbs. in straight line pulling. Weighing approximately 30 lbs. the winch is fitted with 150' 3/16 cable and is compatible with most chainsaws.
The Forest Universe CP 57018-682 PQ Canada G1E 7g3 Tel: 1-877-FOREST 6 email: forest@oricom.ca	**Forestry equipment for ATV's, small and large tractors, skidders, horses, various utility trailers**
Scandinavian Forestry Tech PO Box 427 Kingston, ID 83839 Tel: (208) 682-2002 Fax: (208) 682-2710 email: foresttvone@nidlink.com	**Portable Post Peeling Machine-** small log (1"-10" dia.) peeling machine for tractor PTO. **Small scale logging equipment for tractors and ATV's, forwarding trailers, grapple loaders, cut-to- length processor and delimbers, skidding winches, hand tools, pruning and reforestation tools.**
Wallingford's Inc. 56 Oak St. Oakland, ME 04963 Tel: (207) 465-9575 Fax: (207) 465-9601 email: jjwally@mint.net	**Largest wholesale domestic manufacturer/distributer of logging accessories- including chokers, slings, chains and hardware.**

*****

Appendix II: Residual Damage Measurement Worksheet[1]

1) Prepare a map of the harvest area, then calculate approximate acreage.
2) *(Acres of your harvest operation ___ x.01 = ____ x43560 = ____/10 = _____ The number of feet needed for a 1% cruise line)*
3) Mark approximate location of major skid trails.
4) Lay out cruise lines so they travel *across* trails as much as possible.
5) Plan 10-foot-wide strips to cover 1% of harvest area, or at least 100 trees.
6) Using a hand compass, pacing or hip-chain, measure strips along the planned cruise lines.
7) Use a light stick cut to five feet in length to measure any tree touched within five feet either side of the compass course.
8) Record damage to trees over 5" DBH within the strip on a tally sheet similar to the example below.

EXAMPLE TALLY SHEET

AREA: BROWN WOODLOT
HARVEST AREA: 35 ACRES
MINIMUM CRUISE LINE LENGTH NEED: 1,525 FEET
(1%-0.35 acres X 43,560 sq.ft./acre, divided by 10-foot wide strip)

TREE #	SPECIES	DBH	DAMAGE CATEGORY
1	MAPLE	12"	1 - NO VISIBLE DAMAGE
2	SPRUCE	15"	2 - LIGHT DAMAGE
3	FIR	10"	3 - MODERATE DAMAGE
4	BIRCH	11"	4 - SEVERE DAMAGE
5			
6			

Another example, using tree species abbreviations and short-hand notations:

1. RM-12-1 (**R**ed **M**aple, 12" DBH, Damage Category- 1)
2. RS-15-2 (**R**ed **S**pruce, 15" DBH, Damage Category - 2)
3. BF-10-3 (**B**alsam **F**ir, 10" DBH, Damage Category - 3)
4. WB-11-4 (**W**hite **B**irch, 11" DBH, Damage Category - 4)

Any piece of paper can serve as a tally sheet. Use a clipboard.

---

[1] From *Professional Mechanical Harvesting Practices.* American Pulpwood Association, Inc. Sept. 1997

Appendix III: Logging Cost Calculating for Low-Impact Forestry

I. Introduction

As a logger, no matter how you are paid (by the cord, by the hour, by the day, or even by the acre) you need to know your costs per unit. This information not only helps you come up with a realistic bid for services--allowing you to pay for your equipment as well as yourself-- it also helps you make better business decisions. Careful record keeping can tell you how much productive time, how much downtime, and how much operating expenses go along with your equipment. With this information you can decide whether:

- your equipment is suitable for a given operation,
- it is time to replace your equipment with something more appropriate and efficient,
- you can make a lower bid,
- you should ask for more money per unit, or
- you should reject a potential job.

This report can help you calculate cost/unit if you have good records, or estimate cost/unit if you don't. It also has tables to help you estimate potential logging costs for a specific woodlot given various factors that might increase or decrease your costs. These cost modifiers are important, because with low-impact forestry you will sometimes be asked to cut low volumes of poor-quality wood and take special care to avoid damage to residual trees. Such logging would lead to higher costs per unit of wood removed. Loggers should not be penalized for doing a better job—they should be rewarded! Pricing logging operations correctly is the best way to ensure that the landowner and logger are both getting a fair deal.

The figures gained from the cost calculator can not be considered to be exact. The economics of logging are too complicated for such a result. The figures should, however, be somewhere in the "ballpark" and thus should be helpful in negotiations with loggers and landowners. The cost modifiers, especially, can benefit from feedback from loggers to make them more reflect the realities of logging in Maine. This is thus a working draft that will improve over time.

Calculating cost per unit

If you have been keeping careful records of your logging costs, you can make fairly accurate estimates of your average cost per unit. Such records allow you to view the cost per unit of all the individual cost items, from depreciation to oil changes. Given such detailed information, you might want to make improvements in either equipment or practices to improve your efficiency and returns.

To calculate your costs, list the various cost items by category—fixed or variable. Fixed costs are the costs you would have to pay regardless of how many hours you work your machinery. These costs include depreciation, insurance, interest, and taxes. Variable costs depend on machinery use and include fuel, lubricants, repairs, parts, and purchased services. Horse loggers would have feed, shoes, medicine, and veterinarian costs.

Cost per unit is calculated by adding up the costs of these items per year, and then dividing by either the number of hours of operation per year, or number of cords cut per year.

Fixed costs

Fixed costs include depreciation plus interest, insurance, and taxes. *Depreciation* is normally calculated as the *purchase price* minus the *salvage value* of the machine when its productive days are over (often estimated as 25% of purchase price) divided by the number of years of useful expected operation.

Some loggers, who have paid for their equipment in cash, are confused about *interest*, which is normally the rate of money paid for a loan. Even if you did not borrow to buy a machine, you should still calculate an interest rate—because the money you invested in the machine could have been invested elsewhere and gotten an annual return. If you did not take out a loan, your interest rate is estimated from the return rate you could have gotten from an alternative investment.

To calculate fixed costs you need to provide figures for the following:

Table 1. Estimating Fixed Costs

Symbol	Description	Amount
P	Cost of machinery in dollars	
N	Economic life of machine in years	
S	Salvage value of machine in dollars [0.25 (P)]	
D	Depreciation [(P - S)/N]	
AYI	Average yearly investment [[(P - S) (N + 1) / 2N] + S]	
I	Interest rate (or total of annual payments for machine)	
Ins	Insurance rate (or total of annual payments)	
T	Tax (property) rate (or total of annual payments)	
IIT	Interest, Insurance and Taxes [(I% + Ins% + Tax%)*AYI/100] (or total of annual payments of I+Ins+T)	
AFC	Annual fixed costs [D + IIT]	
PH	Productive hours worked per year	
HFC	Hourly fixed cost [(D + IIT) / PH]	
TV	Total volume (in cords or MMBF) cut per year	
VFC	Volume fixed cost [(D + IIT) / TV]	

Operating costs

The variable costs are the operating costs of the machinery and include costs of *maintenance and repair, fuel, lubricants, and tires and chains.*

The annual cost of maintenance repair (AMR) is determined, if experience and cost data are available, by adding costs of parts and labor to repair the machine (not including oil changes, lubrication and filter costs, and tires and chains, which are covered below) over the economic life of the machinery and dividing by N (economic life of the machinery). Data for one year, though useful, will change as the machinery gets older.

AMR	(annual repair costs)Expected parts and labor over life of machine/N (productive life of machine)	

If experience and data are not available, you will have to use a rough estimate or rule of thumb for a given type of equipment. Research from Canada suggests that these rules-of-thumb formulae are not very reliable.

The formula for estimating maintenance and repair costs is:

AMR	Maintenance and Repair Cost [D * percent rate (see below)]	

The percent rate multipliers depends on type of machine:

Table 2. Percentage multipliers per machine type

Machines	Percent	
	New	Older
Agricultural wheel tractor with winch	25	55
Crawler tractor with winch	30	60
Rubber-tired cable skidder with winch	30	55
Small shortwood forwarder	30	55

Annual *Fuel* costs (F) is determined by multiplying number of gallons used in a year times price per gallon. Annual fuel costs can be estimated by figuring the gallons consumed per hour times price per gallon times number of hours of expected work per year (PH).

F	Fuel Cost [gallons per hour * hours per year * price per gallon]	

Lubricants include: engine oil, transmission oil, final drive, hydraulic oil, grease, and filters) used over a year. To estimate annual lubricant costs (L):

L	Lubricant cost [((oil capacity (gals.) x price/gal) + filter cost) x (PH / hours between oil changes)] +[grease required/yr. (lb.) x price/lb.]	

Other includes fuel and air filters, which can be estimated based productive hours worked per year divided by number of hours between changes of the filters times the cost of the filters.

O	(gas and air filters)PH/ Hours between change x price of filter	

Tires (or tracks) are sometimes subtracted from the initial price of the investment because tires have a shorter life than the machines. With cost data:

T	Annual Tire cost total tire cost in economic life of machine / N To estimate, T = (1 + 0.15) x tire purchase cost x PH/ tire life (in hours)	

The same kind of calculation can be done for tire chains:

C	(chains)C = cost of chains x PH/ chain life (in hours)	

The *total costs for running your machinery* are:

HOC	Hourly Operating Cost [(AMR + L + F + T + O + T) / PH]	
VOC	Volume Operating Cost [(AMR + L + F + T) / TV]	
TAC	Total Annual Cost [Fixed Costs + Operating Costs]	
THC	Total Hourly Cost [TAC / PH]	
TVC	Total Volume Cost [TAC / TV]	

You need to run these calculations for each machine used on a particular logging job, including the *chainsaw*. The working life of a chainsaw is relatively short, compared to other machines. You need to account for purchase price (usually depreciated in 1 or two years), costs of maintenance (chains, bars) repair (parts and labor), and fuel and bar oil. Figure the cost per year and then divide by total volume cut (for cost per cord) or total hours worked (for cost per hour).

Chainsaw costs:

FC	Fixed Cost (purchase price / N (economic life of machine))	
OC	Operating Cost (AMR + fuel + barchain oil)	
AC	Annual Cost (FC + OC)	
HC	Hourly Cost (AC / PH)	

VC	Volume Cost (AC / TV (cords cut per year))	

Total machinery cost per hour or per cord is calculated by adding costs for all machines used.
Total machinery cost:

MHC	Combined hourly costs of chainsaw, skidder, forwarder etc.	
MVC	Combined per cord costs of chainsaw, skidder, forwarder, etc.	

Calculating labor

Whether you are paid by the cord or by the hour, you should make sure that you make a decent wage. If you are paid by the hour, you simply add your hourly costs to total hourly cost to determine the cost for you to log. If you are paid by volume (by the cord, for example), you can determine your hourly wage by subtracting total annual cost from total revenues and dividing by PH (total operating hours per year). For some loggers, the actual labor revenues can be shocking—if they discover they are working to support a machine, but not themselves.

LCH	Labor Cost per Hour	
LCV	Labor Cost per Volume (LCH x PH / TV)	

Base Cost

Your base cost (which is your average for the year) is determined by adding your machinery cost to you labor cost.

BCH	Base Cost per Hour (MCH + LCH)	
BCV	Base Cost per Volume (MCV + LCV)	

Cost modifiers

Once you have determined an average base cost for your machinery, you can use cost modifiers to estimate how much a given operation will cost depending on average tree density, average tree size, difficulty of terrain, or yarding distance.

Cost modifiers will depend on the machinery you use. The following are modifiers for skidders based on figures from a report from Idaho.

Table 3. Felling Cost Modifiers—TDF (Tree Density for Felling)

Original Density (trees/acre)	Average Tree Spacing	25%	% Removed 40%	100%
200	15 ft.	1.00	.90	.90

300	12 ft.	1.08	.93	.90
400	10 ft.	1.11	1.02	.90
500	9.3 ft.	1.14	1.12	.90

Table 4. Felling Cost Modifier--Average Tree Diameter)

Average Tree Diameter in inches	Cost Modifier
6	1.3
8	1.2
9	1.0
10	0.7
12	0.5

Table 5. Ground Skidding Cost Modifiers - Removed Tree Density (TD)

Removed Trees per Acre	Cost Modifier
500	0.80
400	0.80
300	0.90
200	1.00
100	1.13
50	1.37

Table 6. Skid Distance (SD)

Average Skid Distance (feet)	Cost Modifier
100	0.44
300	0.72
500	1.00
700	1.28

900	1.56

Table 7. Difficulty of Terrain (Slope, Roughness, or Deep Snow)

	Favorable Slope	Average Conditions	Adverse Conditions
Cost Modifier	0.9	1.0	1.2

Final calculations

BCH	Base cost per hour [Hourly fixed cost + Hourly operating cost]	
ECH	Estimated cost/hour of using equipment [BCH x felling and skidding modifiers]	
THC	Total hourly cost [ECH + Hourly labor cost]	
TCL By hour	Total cost of logging job [THC x Hours of operation]	
TCL By cord	Total cost of logging job [Cords to be cut x (fixed cost per cord + operating cost per cord + labor cost per cord) x felling and skidding modifiers]	
TC	Total cost for operation [TCL + cost of trail and road building + bridges and culverts and other BMPs + putting operation to bed + cost of hauling equipment	

Note: This cost calculator could use feedback from those who have been keeping good records over time. How close do the estimators come to actual figures? How useful are the modifiers for estimating actual costs? Are there ways that this should be modified to be more practical or accurate? Send comments to publisher.

Sources used to create cost calculator

Some of the formulae used in the cost calculator came from the sources below. Some of the cost modifiers, however are estimates and need further testing.

Frederick Cubbage, W. Dale Greene, and John Lyon, Tree Size and Species, Stand Volume, and Tract Size: Effects on Southern Harvesting Costs, *SIAF*, Vol. 13, 1989.

Harry W. Lee and Leornard R. Johnson, *Calculation Timber Removal Costs Under Ecosystem Management*, Bulletin No. 62, Idaho Forest, Wildlife and Range Experiment Station, Moscow, ID.

Edwin S. Miyata, *How to Calculate Costs of Operating Logging Equipment*, North Central Forest Experiment Station, USDA Forest Service.

Dennis Werblow and Frederick Cubbage, "Forest Harvesting Equipment Ownership and Operating Costs, in 1984, SIAF, vol. 10, 1986.

W.A. Williams, *Predicting Maintenance and Repair Costs of Woodlands Machinery*, Technical Note TN-142, Forest Engineering Research Institute of Canada, Vancouver, BC, December, 1989.

Appendix IV: The Brooks and Neil Investment Chart

Using the chart

The following spread sheet (next page) for two 10 ft. logs with 4 clear faces is an illustration of the BAN-TIC:

DBH is the diameter outside bark at breast height (4.5 ft.).

DBHIB is diameter inside bark at breast.

SCAL.DIB or scaling dib is the top end diameter inside bark rounded down to the nearest inch. **LOG VOL** is the log volume (based on International -inch log rule).

LOG VALUE is log volume times the price and grade.

TOTAL TREE VALUE is the sum of the values of two 10-foot logs.

ANNUAL RATE OF RETURN is computed based on total tree value change from two inches of growth with rate of growth (3, 5, 7, or 10 years to grow two inches) factored in.

RATE OF RETURN CALC calculates rate of return over a given time period.

BROOKS AND NEIL TREE INVESTMENT CHART

Species: Sugar maple

LOG PRICES ($/Mbf)				LOG PRICES ($/Mbf)			
DIB	4CF	3CF	2CF	DIB	4CF	3CF	2CF
7	$0	$0	$0	17	$1,800	$700	$200
8	$0	$0	$0	18	$2000	$700	$200
9	$200	$100	$100	19	$2,500	$700	$200
10	$200	$200	$200	20	$3,000	$700	$200
11	$200	$200	$200	21	$3,000	$700	$200
12	$1.100	$450	$200	22	$3,000	$700	$200
13	$1,100	$450	$200	23	$3,000	$700	$200
14	$1,500	$450	$200	24	$3,000	$700	$200
15	$1,500	$450	$200	25	$3,000	$700	$200
16	$1,800	$700	$200	25+	$3,000	$700	$200

Source of log prices: Floyd's Mill 9/18/98

Log Rule International 1/4 inch Volume Table for 10ft Logs

DIB	Bd Ft.	DIB	Bd Ft.
7	15	17	125
8	20	18	140
9	30	19	155
10	35	20	175
11	45	21	195
12	55	22	215
13	70	23	235
14	80	24	255
15	95	25	280
16	110	26	305

Rate of return Calc. YRS= 10 Begin Value=$138 End Value = $318 ROR = 8.7%

Butt Log (10Ft.) 2nd Log (10Ft.) - 4CF

DBH	DBHIB	SCAL DIB	LOG VOL	LOG VALUE	SCAL DIB	LOG VOL	LOG VALUE	TOTAL TREE VALUE
10	8.5	8	20	$0	7	15	$0	$0
12	10.5	9	30	$6	9	30	$6	$12
14	12.5	11	45	$9	10	35	$7	$16
16	14.5	13	70	$77	12	55	$61	$138
18	16	14	80	$120	13	70	$77	$197
20	18	16	110	$198	14	80	$120	$318
22	20	18	140	$280	16	110	$198	$478
24	22	20	175	$525	17	125	$225	$750
26	24	21	195	$585	19	155	$388	$973
28	26	23	235	$705	20	175	$525	$1,230
30	26	25	280	$840	22	215	$645	$1,485

DBH Years to Grow 2 inches DBH

DBH	3	5	7	10
	ANNUAL RATE OF RETURN			
14	10.1%	5.9%	4.2%	2.9%
16	104.8%	53.8%	36.0%	24.0%
18	12.7%	7.5%	5.3%	3.7%
20	17.3%	10.1%	7.1%	4.9%
22	14.6%	8.5%	6.0%	4.2%
24	16.2%	9.4%	6.6%	4.6%
26	9.0%	5.3%	3.8%	2.6%
28	8.1%	4.8%	3.4%	2.4%
30	6.5%	3.8%	2.7%	1.9%

Butt Log (10Ft.) -4CF 2nd Log (10 Ft.) - 3CF

DBH	DBHIB	SCAL DIB	LOG VOL	LOG VALUE	SCAL DIB	LOG VOL	LOG VALUE	TOTAL TREE VALUE
10	8.5	8	20	$0	7	15	$0	$0
12	10.5	9	30	$3	9	30	$3	$9
14	12.5	11	45	$9	10	35	$7	$16
16	14.5	13	70	$77	12	55	$25	$102
18	16	14	80	$120	13	70	$32	$152
20	18	16	110	$198	14	80	$36	$234
22	20	18	140	$280	16	110	$77	$357
24	22	20	175	$525	17	125	$88	$613
26	24	21	195	$585	19	155	$109	$694
28	26	23	235	$700	20	175	$123	$828
30	28	25	280	$840	22	215	$151	$991

DBH Years to Grow 2 inches DBH

DBH	3	5	7	10
	ANNUAL RATE OF RETURN			
14	21.1%	12.2%	8.6%	5.9%
16	85.3%	44.8%	30.2%	20.3%
18	14.2%	8.3%	5.9%	4.1%
20	15.6%	9.1%	6.4%	4.4%
22	15.1%	8.8%	6.2%	4.3%
24	19.7%	11.4%	8.0%	5.5%
26	4.2%	2.5%	1.8%	1.2%
28	6.1%	3.5%	2.6%	1.8%
30	6.2%	3.7%	2.6%	1.8%

Appendix V: Sample Contract

(from Maine Forest Service)

This agreement is made this ____________ day of _____________, ________ by and between:

Seller:

Name(s): ______________________

____

____

Address: ______________________

____

____

Tel # ______________________

SS # or Federal ID # ______________

____

Purchaser:

Name(s): ______________________

____

____

Address: ______________________

____

____

Tel # ______________________

SS # or Federal ID # ______________

____

(This Agreement refers to "Seller" and "Purchaser" throughout. Where more than one Seller or Purchaser is a party to this Agreement, references to "Seller" or "Purchaser", as applicable, shall mean all Purchasers or Sellers collectively).

The Seller agrees to sell to Purchaser and to allow Purchasers entry upon Sellers' land, upon the terms and conditions stated below:

I. Property Location/Access/Boundaries

Seller grants to Buyer permission to enter Seller's land, together with workers and equipment upon the terms and conditions of this Agreement, to harvest forest products from the areas designated by Seller and remove the forest products listed in this Agreement. Buyer agrees to cut and remove the forest products and to pay Seller according to the terms of this Agreement. (Seller to determine adequate level of detail).

A. Lot Location and Description

Seller's land ("the Lot") subject to this Agreement is located in (town/township/plantation) ______________________, ____________________ County, Maine. The Lot is known as the ______________________ lot as more fully described in (identify deed or other documented source of title):

____

____

Designation of Area to be Cut/Survey

The area to be cut contains approximately ______________________ acres.

Check appropriate provision

____________ The entire Lot is subject to the terms of this Agreement.

____________ Only a portion of the Lot will be subject to harvest operations.

The areas upon which Purchaser may enter and cut are depicted on (identify sketch or a survey provided by purchaser to Seller) ____
dated ____________________________, prepared by____________________________________.

Other description of area to be cut if no sketch or survey provided____
____
____

B. **Marking Boundaries**

On land where 10 acres or more is to be harvested, the property owner is by law responsible for clearly marking the property line if the cutting is being done within 200 feet of the property line (Title 14 MRSA ss. 7552A). With respect to this requirement, the parties agree that the responsibility to mark boundary lines, regardless of the acreage of the area to be harvested, shall be met as follows:

Check appropriate provision

________________ Seller agrees to be responsible for marking property lines prior to Purchaser's cutting.

________________ Purchaser agrees to determine and mark the property lines at Purchaser's costs before commencing harvesting activities.

If applicable, Purchaser, in marking the boundaries is relying on the following information provided by Seller: (Identify plans or survey information).
____
____
____

Less than ten (10) acres will be harvested and/or limits of the cutting area are sufficiently within the interior of the lot to be more than 200 feet from the nearest property line. Therefore, neither Seller nor Purchaser is obligated under this Agreement to mark boundary lines. Notwithstanding the parties' waiver of any survey requirements under this Agreement, Purchaser shall be responsible for overseeing the cutting operation to ensure that cutting occurs only in the designated areas, and that timber trespass is avoided.

Trees on the boundary line with adjacent landowners shall not be cut.

Trees cut by Purchaser on land of Seller outside areas designated for harvest shall be purchased at (specify ration) _______________ times the price otherwise applicable under this Agreement.

C. **Access**

Check appropriate provision:

_______________ Access will be provided by Seller as indicated below.

_______________ Access will be arranged by Purchaser as indicated below.

1. **Access Provided by Seller**

Access from the nearest public way (add name of publicway)________________________ ____ to the designated cutting areas shall be over and upon the Lot, unless stated otherwise below.

Description of access provided by Seller, if other than over Seller's Lot from a public way: (Identify title document, license or other source of Seller's right of access).

____

____

2. **Access Provided by Purchaser**

Check if appropriate.

________________ Seller does not have legal access to the area to be cut. Purchaser shall be responsible for obtaining such access at Purchaser's cost. Harvesting operations will not proceed until Purchaser has obtained all necessary licenses, permits, or other legally binding permissions from other landowners to travel over their land.

Purchasers Use of Access

As applicable, Purchaser agrees to abide by the terms, rules and regulations governing Seller's or Purchaser's rights of access to the Lot.

D. **Seller's Warranty of Title**

Seller is the owner of the Lot and the timber on the Lot with the full authority to sell the timber under the terms of this Agreement. If Seller has designated or obtained access rights over lands of others to be used by Purchaser, Seller hereby assures Purchaser that Purchaser may exercise such rights of access without further grant or permission from the other landowners.

II. **Term**

Purchaser shall commence harvesting on ______________________________ and shall complete harvesting by ______________________________, unless this contract is terminated as elsewhere provided in this Agreement, or the contract is extended by the parties in writing.

III. **Description of Timber to be Cut and Removed**

Buyer shall remove and pay for the forest products described on **Appendix A.**

Payment shall be made (specify weekly, monthly, or other arrangement)__________________
____ at Seller's address specified above for forest products removed, as scaled or measured under the terms below. Legible scale slips shall accompany the payment.

IV. **Status of Parties**

A. **Designated Forester or Agent**

For purposes of oversight of Purchaser's compliance with this Agreement, in addition to review by Seller, the parties agree that the person named below (if any) shall also be deemed the designated agent of Seller:

Name of Seller's forester/agent:
____

Forester/agent'saddress:____
____

Telephone Number:____

If Agent is a Maine licensed professional forester, state license number:____________________

The Seller's forester/agent shall be the agent of Seller with authority to review and approve forestry activities
on the land during the term of this Agreement, and Purchaser agrees to consult with the forester/agent and abide by the forester's/agent's determinations and instructions to the purchaser during all stages of the harvest under this Agreement.

B. **Purchaser's Status; Purchaser Responsibilities and Warranties:**

Notwithstanding any other provisions of this Agreement, no relationship of employer/employee or master/servant between the Seller and the Purchaser or between the Seller and any agent, employee or subcontractor of the Purchaser shall be deemed to exist. Purchaser shall select its own employees, set rates of pay and all terms and conditions for employment, and pay Purchaser's own employees, agents or subcontractors. Neither the Purchaser nor its employees agents or subcontractors shall be subject to any orders, selection, supervision or control of the Seller. It is mutually understood and agreed that the Purchaser is deemed to be an independent contractor. Nothing herein contained shall prohibit the Purchaser from contracting to purchase and harvest forest products on land of others.

Purchaser warrants and represents that Purchaser does and will employ and utilize the equipment and personnel necessary to perform the harvesting contemplated under this Agreement in a timely manner. Purchaser shall be solely responsible for the acquisition, maintenance, replacement and repair of its equipment, and for the selection, training, supervision, control, direction, compensation, work rules, discipline and termination of its employees or subcontractors. Purchaser warrants and represents that all of its employees will perform in accordance with the requirements of this Agreement when assigned to the work to be performed hereunder. Purchaser will equip and train its employees and subcontractors adequately to perform the required services in a safe, timely and lawful manner.

Purchaser will conduct Purchaser's business to be at all times in full compliance with all requirements of Federal, State, and local law, including applicable common law, statutes and requirements, and including but not limited to the requirements of the federal Fair Labor Standards Act, all federal and State labor and employment laws, federal immigration laws, the workers' compensation laws, federal and state equal employment laws, the Internal Revenue Code and State tax laws and regulations, the unemployment insurance laws, the federal Occupational Safety and Health act of 1970, as amended, and its regulations, state laws pertaining to occupational safety and health, state laws and regulations pertaining to wood harvesting, and any other laws or governmental rules and regulations pertaining to the services to be provided hereunder.

The Purchaser will ensure that full timely payment is made:

1. for all employee wages and benefits, fuel and supplies;
2. for the lawful disposal of any regulated or hazardous waste or substances it handles;
3. of any and all contributions or taxes for unemployment insurance, old age retirement benefits, Workers' Compensation or any other such employee entitlements now or hereafter imposed by law.

Purchaser is and will remain in compliance with the Maine Workers' Compensation Act and Maine Employment Security Law. Purchaser agrees to indemnify the Seller from all loss, cost or expense, including defense costs and attorneys fees, arising by reason of the breach of any of these warranties or representation. (Other provisions, if applicable)

____

____

C. Subcontractors and Employees

Purchaser may not contract with a third party to perform any part of the harvest operations contemplated under this Agreement without the written consent of Seller. All subcontractors shall be deemed agents of Purchaser for purposes of this Agreement.

The independent contractor will not hire any employees to assist in the wood harvesting without first providing the required certificate of insurance to the landowner.[1]

[1] This statement is required by law if the landowner seeks to file an "Application for Predetermination of

V. **Forestry Practices**

The following are minimum forestry practices applicable to this Agreement. Purchaser will, at Purchaser's sole cost and expense, harvest the designated types of species of wood from the designated cutting areas during the terms of this Agreement in accordance with the accepted principles of professional forestry, the Maine Forest Practices Act and rules and regulations promulgated under 12 MRSA, Chapter 805, Subchapter III-A, and the following agreed standards of performance.

A. **Harvest Notification**

Before Purchaser begins harvesting operations, the party designated below shall notify the Maine Bureau of Forestry, as required under Title 12, ss. 8883 of the Maine Revised Statues.

( ) Seller
( ) Purchaser
( ) Other designated agent (specify)______________________________

The party indicated above shall retain a copy of the notification form and, unless otherwise specified below, shall be responsible for reporting harvest information in compliance with Maine law. If Purchaser or a designated agent other than Seller has the responsibility of filing such harvest reports, the person responsible for such filings shall provide copies of the reports to Seller at the time they are submitted to the Maine Forest Service.

B. **Scaling**

All wood meeting the specifications of the parties as set forth on Paragraph III of this Agreement shall be
measured as specified below.

Volume or Weight Sales

Check appropriate item:

____________ Sales of volume shall be measured in standard cords, board feet, tons, or pounds in accordance with the Wood Measurement rules.

____________ The parties agree that weight measurement, in accordance with the Wood Measurement rules, may be used.

Independent Contractor Status to Establish Conclusive Presumption" with the Workers' Compensation Board.

Sale of Tree Length Wood

Check appropriate item:

_____________ Butt measure shall be the standard method for measurement of tree length wood purchased under this Agreement.

_____________ The parties agree that the following method of measurement may be used, rather than butt measure, for tree length wood.____________________________

____

____

____

Log Length Stems (partial bole sections of tree length material)

Log length shall be measured as follows:

_____________ International ¼ Inch Log rule shall apply.

_____________ Other (identify)____

Scaling of products, including scaling procedures and scaling records, shall be carried out and maintained in accordance with the directions of Seller. In the event that scaling is done on the Lot at harvesting site, it shall be done by a person or persons acceptable to Seller (who shall in any event be State licensed scalers) and the cost of scaling shall be paid by Purchaser.

In the event that scaling of products is done off premises, it shall be done in a mill yard or at such place as is acceptable to Seller.

Reports of volume (legible stumpage sheets, measurement tally sheets or the like) shall be provided in full to Seller on a (weekly) (bi-weekly) or (monthly) basis by Buyer as wood is delivered to receiving mill.

Further, Purchaser shall forward a (weekly) (bi-weekly) or (monthly) report showing in full the volumes for all wood products hauled from the Lot. Such volume reports will include the following:

___________ the name of the harvest contractor or subcontractor
___________ date and time of loading
___________ product type and species
___________ mill of destination
___________ the name of the hauler

C. **Utilization Requirements**

1. Harvesting shall proceed in an orderly manner from the back of the lot to the front, or in an equivalent manner which will ensure completion of cutting in all areas designated for harvest. Only wood designated by the Seller/Seller's agent shall be harvested.

2. All marked or otherwise designated wood, including defective trees, must be cut.

3. Stump heights shall not exceed six (6) inches, except where obvious obstacles, problems with terrain, swell of roots, or similar hindrances do not permit such a low cut. Snow shall be removed as necessary to comply with this requirement.

4. All trees are to be limbed and cut off at the top end so that no part of any felled trees which will make merchantable product shall be left in the woods unutilized.

5. Small end diameter shall not be less than the minimums specified below for the indicated types of wood and product.

____

____

____

6. **Skidding**

Outside of areas designated for clearcutting and landings, insofar as ground conditions permit, trees shall not be skidded against residual trees or trees for reproduction.

D. **Condition of Roads**

Purchaser agrees, at its expense, to construct roads, and skidder trails in accordance with the appropriate rules of the Maine Land Use Regulation Commission and/or the Department of Environmental Protection, and any applicable municipal ordinances.

Purchaser agrees to maintain and leave any existing access roads in the same or better condition than when harvesting began. The cleared size of landings shall not exceed that needed for safe and efficient skidding and loading operations.

E. **Transportation Facilities**

Purchaser may construct and maintain roads, bridges and other access appurtenances as needed for harvesting. The location and clearing widths of all haul roads and landings constructed by Purchaser shall be agreed to between Purchaser and Seller. Such Agreement shall be by written memorandum before construction is started.

Check appropriate provision:

__________ Purchaser is authorized to cut and use timber for construction, without charges, for forest product transportation facilities located on Seller's Lot. Unmerchantable timber shall be used for such facilities to the extent practicable.

__________Timber cut from the Lot by Purchaser and used in construction of transportation facilities, including road construction, shall be paid for by Purchaser at the rates applicable under this Agreement.

F. **Slash**

Purchaser shall be responsible for disposing of all slash resulting from harvesting operations, so that none shall remain on the ground within twenty-five (25) feet of the adjoining property lines. For purposes of this paragraph, adjoining property lines shall include, in addition to land of third parties, the boundaries of railroad rights of way, and electric power, telephone, pipeline and other utility easements. Purchaser shall also remove all slash a distance of fifty (50) feet from the bounds of any adjoining highways or public ways.

Purchaser shall not place, deposit or discharge, directly or indirectly into any inland or tidal waters, or on the ice or banks of such waters, any materials resulting from the harvest of forest products (including slabs, edgings, sawdust, shavings, chips, bark or other forest products refuse) in such a manner that they may fall or be washed into such waters or in a manner which would allow drainage from such deposits to flow or leach into such waters.

G. **Litter/Pollution Avoidance**

Purchaser shall not discard or otherwise dispose of litter on the property of Seller or any private property, into waters of the State or on ice of such waters, or upon any adjacent highway or public way, and shall be responsible for off site disposal of garbage and refuse generated by forest operations in a lawful manner. For purposes of this paragraph, litter means all waste materials, including bottles, cans, machine parts and equipment, junk, paper, garbage and similar refuse, but shall not include the wastes of the primary processes of forest product harvesting, such as sawdust and slash.

Purchaser shall not service skidders, trucks or other equipment at locations where pollution of waters of the State of Maine is likely to occur.

H. **Fire Suppression**

Purchaser shall comply with all forest fire suppression laws of the State of Maine.

I. **<u>General Compliance with Forestry, Land Use and Environmental Laws</u>**

Without limiting the scope of the preceding paragraphs, Purchaser shall comply with all laws, ordinances and regulations of the municipality where the Lot is located (if the township is organized), the State of Maine and of the United States relating to timber cutting; removal and disposal of slash, debris and litter; construction of roads, trails and landings; protection of streams, rivers and other waters of the State of Maine; soil erosion; and all other laws, regulations and ordinances pertaining to forest product harvest operations and their effect on the environment and land use, including but not limited to, the applicable standards of the Maine Land Use Regulatory Commission and rules and regulations established thereby and forest regeneration and clear-cutting standards of the Bureau of Forestry, Department of Conservation of the State of Maine adopted under the Maine Forest Practices Act. Best Management Practices as published in <u>Erosion and Sedimentation Handbook for Maine Timber Harvesting Operations</u> will be implemented.

Purchaser warrants that Purchaser will promptly notify Seller on any occasion on which Purchaser may be cited for a violation of laws governing the harvest operation.

VI. **<u>Default/Enforcement of Obligations</u>**

Upon the occurrence of any event of default by Purchaser, Seller may, at any time thereafter, do any or all or any combination of the following:

A. Seller reserves the right, for good cause, to halt Purchaser's harvest operations and terminate this Agreement, if in the opinion of Seller or Seller's Designated Forester/Agent, the Purchaser is breaching the terms and conditions of this Agreement.

B. Enter into the Lot and take possession of all forest products remaining on the Lot.

C. Require Purchaser to give an accounting of all forest products hauled from the Lot or yarded thereon.

D. Require Purchaser to pay stumpage at rates and scales specified in this Agreement for all merchantable material left in the wood or wasted in stumps or tops.

E. To grant other permits to third parties to complete the harvesting specified in this Agreement in the event of termination of this Agreement or for unexcused harvesting stumpage by Purchaser.

F. Take corrective action as Seller deems necessary to abate erosion or damage to the Lot and to remove slash, litter and abandoned property of the Purchaser, at Purchaser's cost.

G. Enjoin any activity of Purchaser in default of this Agreement, and/or seek any other judicial or administrative remedy available to Seller at law or in equity.

Upon the termination or completion of this Agreement, Seller or Agent may examine the Lot and any access road and report to Purchaser any failure on the part of Purchaser to comply with the conditions, terms and specifications of this Agreement.

VII. **Insurance**

Purchaser shall provide and maintain during the term of this Agreement insurance as follows:

A. **Workers' Compensation and Employer's Liability Insurance:**

1. Purchaser shall take out and maintain during the term of this Agreement, Workers' Compensation Insurance covering all its employees and any others performing work related to this Agreement, with the coverage set forth in Maine Statutes, and Employer's Liability Insurance covering all such persons; or

2. A notarized statement to the Seller that he/she is an independent contractor (will not employ other workers during the harvest outlined by this agreement). The Seller will obtain a declaration of independent status of the purchaser from the Workers' Compensation Board.

B. **Public Liability and Property Damage Insurance:**

The Purchaser shall take out and maintain during the term of this Agreement, Public Liability and Property Damage Insurance to protect against claims for damages for bodily injury, including personal injury to or destruction of property which may arise from operations performed under this Agreement. The minimum amounts of such insurance shall be as follows:

Bodily Injury Liability $100,000 each person, unless another amount is specified here: $ _______________

$500,000 each occurrence, unless another amount is specified here: $ _______________

Property Damage Liability $100,000 each occurrence, unless another amount is specified here: $ _______________

VIII. **Indemnity**

Purchaser shall indemnify and hold Seller and Seller's forester, agents, and employees harmless from and against any and all manner of claims, suits, fines, penalties and expenses incurred by Seller, and/or Seller's forester agents and employees arising or allegedly arising out of the performance of this Agreement by Purchaser and Purchaser's agents, employees, contractors or invitees or on account of Purchaser's use of the Lot or its access.

In the event Seller shall be forced to resort to legal action to enforce any provision of this Agreement or to defend against claims or actions resulting from Purchaser's performance under this Agreement, Purchaser shall be responsible for all Seller's costs, including reasonable attorney and paralegal fees and court costs, and the cost of any professional services necessary for the determination of fault or the scope of Purchaser's non-compliance with this Agreement. Purchaser's agreement to hold Seller harmless under this paragraph shall survive the termination or expiration of this Agreement.

IX. **Assignment**

Purchaser shall not assign this Agreement without Seller's prior consent in writing.

Entire Agreement

This contract contains the entire agreement of the parties, and neither party shall be bound by any statement or representation not contained in this Agreement. No consent or waiver, express or implied by the Seller to or of any breach of any obligations of Purchaser under this Agreement shall be construed as a consent or waiver to or of any other breach of such obligations. This Agreement may be amended only by a writing signed by the Seller and Purchaser, and deny other person against whom enforcement of this Agreement is sought.

The parties have subscribed their names to this Agreement, agreeing to be bound by it, as of the date stated on the first page of this Agreement.

______________________________ ______________________________

______________________________ ______________________________

SELLER (S)

____

____

____

____

____

____

PURCHASER (S)

Appendix A

Price and Terms of Payment

Purchaser shall pay the following prices for forest products removed.

1. Sawlogs	Species	Grade	Price/MBF

2. Veneer	Species	Grade	Price/MBF

3. Pulpwood	Species	Per Cord	Per Ton

4. Boltwood	Species	Per Cord

5. Other Products	Species/Product	Price	Unit of Measure

6. Chips	Price	Unit of Measure

Appendix VI: Recommendations to FSC-US to Make Forestry Certification More Credible

(written in Winter of 2000-2001)

Introduction. Many environmental groups think that forest certification is the best way to encourage companies to practice good forestry. Certification uses market carrots, rather than regulatory sticks. Of the array of forestry certification programs available, most environmental groups favor the collaborative-based Forest Stewardship Council (FSC), rather than an industry-based system, such as the Sustainable Forestry Initiative (SFI).

Even though FSC standards are derived through lengthy collaborations from a broad spectrum of interest groups, there have been a number of certifications lately--both world-wide and in the U.S. and Canada--that some environmental groups, native groups, and labor groups have protested. These protests seem to be an odd response to practices that are declared to be ecologically sound and socially responsible by certifying experts.

In November of 2000, FSC-US sent out a draft of its National Indicators. I reviewed these indicators, at the request of the Maine Sierra Club (which is appealing a recent FSC certification in the state) to see what was present or missing that would allow certifications that are so controversial. How could standards be changed to lead to more credible results? The following is an edited version of my comments to FSC.

Time. For a forestry certification system to be successful, it must have credible standards that are transparent to both landowners and the public. There should not be a major disconnection between what the landowners are supposed to be doing and what members of the public see if they go on the landowners' operations.

One key factor, that was not emphasized in the latest Indicators Draft, is the concept of time. Some organic farming standards, such as those in Maine, require that the land be managed without chemicals for three years and that there be a certain measurable level of organic matter in the soil before the farm can be certified.

Forests, in contrast, have much longer time horizons than annual crops on farms. One cannot get desirable forest composition or stand structures over a landscape in one, two, or three years. This suggests that there be a two-tiered approach to time:

1. Some standards can be, and thus should be, in practice for a time before certification is granted. This includes such items as: cutting less than growth, minimizing damage in logging, paying loggers a "living wage," or avoiding pesticide use. Indeed, most of the standards, with the below exceptions, should have a history of practice by the landowner (for a minimum time period to be designated by the Regional Standards Committee) before the landownership can be certified.
2. For those elements that would take decades to develop--such as desirable landscape composition, stand structures, or age class balance--the landowner should have management plans in place that would reasonably lead to the desired outcomes.

At the time of certification, if a company, for example:

- is cutting more than growth;
- is relying primarily on logging technologies that leave a big footprint in the woods;
- overworks its foresters so that they have no time to mark trees in partial cuts;

- tends to dominate local economies and sets a regional standard for squeezing logging contractors;
- uses more clearcutting, herbicides, or plantations than the regional average;
- only the year before certification, uses herbicides questionable under FSC standards[1]--

some people would, rightfully, be confused as to the integrity of the process and would wonder what is being certified--practices or promises? Having measurable targets over some reasonable time period would avoid such confusion.

Measurable targets. Some of the draft indicators are unclear as to what particular result is required of the forest manager. The provisions are broadly stated and open to a wide range of interpretation. For example, 5.3.b states that "Harvest is implemented in a way that protects the integrity of the residual stand. Provisions concerning acceptable levels of residual damage are included in operational contracts." In 6.5.d, the standards state that "Logging damage to regeneration and residual trees is minimized during harvest operations." But what is "acceptable" and what is "minimized"? This is up to interpretation of certifiers who may be looking more at local general practices than at measurable, desirable outcomes.

The monitoring section, 8.1, does not even suggest monitoring for these very important outcomes. Monitoring for logging impacts without some sort of measurable target would generate data without specifying any way to determine if the operation is in the ballpark or not.

For all the importance of management plans, it is the actual logging operations that lead to results on the ground that are either examples of a well-managed forests or not. The following are areas where regional committees can set scientifically-based, measurable targets that can be used (with flexibility according to terrain, stand type, and stand conditions) to measure performance:

- *By forest type, and adjusted for terrain, targets for maximum allowable percentage of land put in roads, trails, and yards (or landings).* Roads, yards, and trails not only have impacts on soil and water, they also can: lower productivity, by lowering the percent of land that is actually growing trees; fragment closed-canopy, interior forests; and reduce the number of crop trees suitable for long-term management. Scandinavian countries have measurable standards for this, and there is no reason why FSC cannot do the same.
- *By forest type, targets for reducing level of undesirable soil disturbance over logging operations (such as compaction or rutting).* (For an example of how this can be done, see Chapter 7 of this book)
- *By forest type, targets for percent of moderate to severe logging damage to residual trees and regeneration from a logging operation.* (For an example of how this can be done, see Chapter 5 of this book.) In Sweden, the goal is to reduce such damage to less than 5% of crop trees. Logging damage can affect tree health, and long-term tree quality.
- *By forest type and habitat, stocking targets for partial cuts.* Stocking can influence productivity and quality. Poorly-stocked stands can lead to higher rates of blowdown, undesirable branching in trees, and a shift in regeneration to earlier successional stages. Poor stocking can also degrade a closed-canopy, interior forest habitat. For the northeast, for example, US Forest Service has stocking guides by forest type for A-, B-, and C- line stocking. Stocking guidelines for riparian areas, or other special habitats, might be different than stocking only for productivity (this was recognized by the Maine Council on Sustainable

---

[1] All of these were true at the time of the certification of J.D. Irving's Allagash Timberlands in Maine

Forest Management appointed by the governor). The FSC draft indicators, for some reason, neglected to focus attention on stocking.

- *By forest conditions and management objectives, targets for % marking by foresters (or forest technicians) of areas to be logged.* Some companies leave the decision of what trees to cut up to loggers. These loggers may not be trained as foresters or ecologists to make the best decisions. They might also have an incentive to highgrade. They might, even if well-intentioned, not be able to see all sides of the tree or tree crowns, especially if they are in a machine cab and working on a night shift.

  While in some circumstances, cutting decisions do not require tree marking (clearcuts or overstory removals, for example), for many types of partial cuts, silvicultural and ecological decisions are better left to foresters than to loggers, who may be more concerned with productivity than quality.

- *Targets for acceptability of whole-tree harvesting.* I would not suggest that FSC ban any particular type of machinery, but I see a need for cautions on some types of logging systems that may make meeting FSC goals difficult. In Maine, feller-buncher/ grapple skidder/ delimber systems typically require trails 14 or more feet wide and separated by 40 or so feet. This can put 25% of logging area in trails alone. Removing whole trees (boles with tops and branches still attached) in bunches increases likelihood of damage to residual trees along trails. Yards, for storing whole trees to be delimbed, are often relatively large. Slash, even when it is taken from the yard back to the trails, is not evenly distributed back in the forest from whence it came.

  When a company does the majority of its logging with such a system, this weakens certification credibility. *The claim that other landowners in the region are doing these practices does not justify them as certifiable--if such practices do not meet the ecological or social goals of FSC.*

Measurable targets can also be extended to management plan goals (ones that, as mentioned earlier, take decades to achieve). For example:

- *By forest region, long-term management targets for percent of forest in key ecological conditions important to biodiversity.* Examples of this include closed-canopy, late-successional forests, or old-growth-like structures. The regional committee would have to define their terms to ensure that the results have true ecological integrity.

  While 6.3.a.2, for example, calls for landowners to maintain or restore a range of age-classes, this call has no clear goals attached. Some landowners have made up their own standards defining what is "old" and what is acceptable for the shape or size of such stands. In Maine, for example, some companies think that is acceptable to use riparian zones that are only 250 feet wide to meet these requirements, even though such stands might be considered more "edge" than "interior." Regional committees will have to make an attempt at setting targets, otherwise such indicators will be up to the discretion of cerifiers.

ACE (the allowable cut effect). Indicator 5.6 states that rate of harvest shall not exceed levels which can be permanently sustained. There is a degree of latitude as to how this can be determined. The draft indicators have little to say about companies who are cutting more than average growth rates, based on expectations of higher growth rates in future from intensive management.

The ACE strategy is controversial. Cutting more than growth lowers the immediate inventory. It can, in the short term, lower the percentage of older stands and sawtimber stands, shifting growth to younger age-classes. Basing current cuts on projected future growth assumes that an even-aged system dependent on pre-commercial thinning, herbicides, and short rotations of stands that are relatively uniform (compared to natural stands) will work in reality the way it works in the computer. The stands will not be subject to problems of drought, windthrow, insects, diseases, or lack of adaptation to microsites in the landscape over the coming decades. Such assumptions may not be very realistic.

The intensive management systems also seem to violate the intent of having landownerships that do forestry that is ecologically sound (see following discussion on plantations). These systems also violate indicator 6.6--to use environmentally friendly non-chemical methods of pest management and strive to avoid the use of chemical pesticides. Companies doing intensive management are setting up conditions where pesticide use is more likely over the long term.

Even-aged systems that create large enough openings also create habitat for pioneer species or stump sprouts that landowners might feel obliged to spray with herbicides. In Maine, herbicide spraying and pre-commercial thinning of black spruce stands have made stands more susceptible to the yellow-headed spruce saw fly, which some landowners have sprayed with broad-spectrum chemical pesticides. Using herbicides to encourage fir-dominated overstories can also encourage increased spruce budworm damage.

The term "avoid" can be interpreted in a way that allows a company to spray, for example, more herbicides than the regional average. The company might claim that it is "avoiding unnecessary use." That raises the question of whether any company would intentionally spray money away on an unnecessary use. It does not change the fact that even reducing use from the recent past, the company is still spraying more than its neighbors who are not doing intensive management.

FSC's lack of attention to this controversial issue has meant that certifiers can, and have, certified companies that are cutting more than growth based on ACE and are spraying more herbicides than is average for the state. This behavior has not helped the cause of transparency or credibility.

Plantations. There is a reason why plantations are controversial. Some types of plantations violate basic principles of ecological management. Unfortunately, the FSC draft document does not define "plantation." One company in my region, and its certifier, have claimed that a stand that has been clearcut, sprayed with herbicides, and planted to a single species of tree that would not normally dominate the site is not a "plantation," but a "planted forest." They argue that the tree species planted are either found in the region or are related to trees found in the region and the stands have some of the characteristics of a natural forest. This argument, similar to former President Clinton's argument that what he did with Monica was not "sexual relations," fails the straight-face test and does not help the credibility of certification.

FSC recognizes an array of acceptability of various types of plantations (indicators looking at stand diversity, exotic species, or scale and layout of plantation blocks, for example), but it did not organize this array in a way that enables regional committees to make more clear assessments of acceptability. Certifiers, who have an interest to attract paying clients, can thus come up with their own standards.

Here is an example of how to arrange an array of acceptability for plantations (items under bullets go from less acceptable to more acceptable). FSC should recognize that it is possible to manage natural regeneration into plantation-like stands through use of herbicides, pre-commercial thinning and shortened rotations. Such stands should be judged under the same criteria.

The array recognizes gray areas. It can be turned into a rating system that helps committees decide what is more or less acceptable in planted or intensively-managed stands.

Plantation Array

- Purpose of plantation:

--grow fiber fast;
--fill in gaps;
--restore long-term forest ecosystem.

- Vegetation control:

--broadcast aerial herbicides;
--spot herbicides;
--thinning by hand that allows retention of good examples of all species.

- Species diversity:

--monoculture of exotic species;
--monoculture of a regionally native species that would not normally dominate site;
--monocultureof native species that would normally dominate site;
--diversity of native species adapted to site.

- Stand structure diversity:

--dead standing and down trees removed, slash removed, uniform, even-aged stands;
--some retention of dead and larger living trees, but still tends towards uniformity;
--mimics diversity of natural stand.

- Planned rotations:

--a fraction of biological maturity;
--trees big enough for small sawlogs, but still biologically immature (mostly juvenile wood);
--trees large enough for larger sawlogs, stand develops some older characteristics;
--some trees allowed to get old, stand allowed to develop uneven-aged characteristics.

- Landscape context:

--dominates forest landscape;
--dominates certain forest types;
--intrudes and/or fragments natural interior forest;
--on margins of natural forest and used only rarely;
--(restoration) enhances natural forest.

A plantation to restore a forest on former agricultural land may indeed meet the goal of 10.2 to "promote the protection, restoration, and conservation of natural forests, and not increase pressures on natural forests." Attempts to cut down natural forests and convert to plantations are highly questionable on these grounds, however. The exception would be a highly degraded forest that is planted to restore native species and diversity. If the landowner is clearcutting a natural forest, spraying herbicides, and planting species that would not naturally dominate the site, certification of such practices would lead to legitimization of a type of forestry long protested by local citizens and environmentalists.

It should not be the purpose of FSC to allow landowners to meet ever-rising short-term demand of wood products by allowing intensive management practices that go against ecological principles. Many of these demands are for wasteful and frivolous uses, hardly justifying the sacrifice of forest ecosystems. It might be more fruitful to call on society to reduce such wasteful and frivolous demands, rather than call on landowners to meet them with questionable practices.

No matter how intensively forests are managed, there are limits to what they can produce. Since it is a given that forests have limits, this implies that society will have to eventually live within those limits. It is far better for society to live within limits that are based on sustainable management of more whole ecosystems than on management systems that hurt ecosystem integrity and may be much harder to sustain over many rotations.

Social red flags. While the indicators are, rightfully, stated as positives, there are some social negatives that certifiers might note as "red flags" indicating a need for closer attention. When an independent observer sees a number of these, it is difficult to take certification as being "socially responsible" seriously.

Some companies, due to their size and location, can have a major influence over local economies (monopoly, oligopoly, monopsony, oligopsony) and political systems (donations, lobbying, and use of economic leverage). Companies with this degree of power have an added obligation to use it in constructive ways. Large companies competing in global markets have a temptation to use their clout to reduce costs in ways that can hurt local communities. Certifiers therefore need to ask the following questions:

- *Does the company use this power to depress wages (are wages in this region less than in other forest regions for similar work and is the company setting the industry standard for wages)?*
- *Does the company (if it has mills) use its power to reduce payments to woodlot owners for purchased wood?*
- *Does the company leverage its work force to work long hours or have contractors work equipment on nightshifts?*
- *Does the company make use of imported labor working at wages unacceptable to local workers?*
- *Does the landowner export a large portion of its raw sawlogs to foreign export markets rather than support local mills?*
- *Does the company use its power to influence public policy to assure its ability to lower its taxes (shifting taxes to others), ensure cheap labor, protect against reasonable forest policy changes, or prevent regulations that might protect the environment?*
- *Does the company use its influence over the standards-making process to weaken certification standards and make them more "industry-friendly"?*

Some of the above concerns were mentioned under 4.1, yet there are FSC certified companies in Maine whose practices have raised these red flags for years and have contributed to the decline of local communities. Perhaps FSC needs to find a way to state the positive goals of supporting local economies and paying a fair wage more forcefully so that it is unambiguous to certifiers what this means.

Who are the certifiers? While certification is touted to be an objective process done by third party "independent" certifiers, the reality may be different. Indeed, there are pressures within the certification system that can, if not checked, lead to weakened standards and lax interpretations of these standards.

Some certification companies are for-profit operations that make more money as they certify more acres. FSC itself has the temptation to encourage rapid growth of certification to increase market power, increase prestige of the organization, and increase the power of those within the organization's bureaucracy. To some extent such rapid growth is restrained if the credibility of the organization falls, due to increased controversy.

While FSC is a multi-stakeholder organization, some stakeholders can have a powerful influence due to economic clout. One of these influences is funders. Funding is supposed to be done with no strings attached. In 1998, however, major funders (including Rockefeller Brothers Fund, Global Wallace Foundation, MacArthur Foundation, and Ford Foundation), suggested in a letter (not available to FSC membership) that forest products industry membership be better represented on the FSC board. Such a demand surely sounds like "strings" and does put pressure on FSC to change policies.

The biggest clients of certifiers are large landownerships--indeed 96% of land certified worldwide has been industrial and governmental holdings. These large landholdings give an immediate huge boost in certified acreage. In Maine, just two landowners under certification (or in the process of being certified) represent nearly 2.5 million acres of certified land (1.5 million acres of these lands have already been certified).

Some landowners being certified have representatives on the regional committees that make the regional standards. To the extent that the committees operate by consensus, these companies have veto power over standards that might restrict their certifiability. One of these landowners, who owns land in neighboring New Brunswick, dropped its Canadian certification when the regional committee passed standards to which it objected concerning the use of pesticides and exotic species. This landowner objected that there was not enough industrial representation on the standards committee.

Although ecosystems do not operate by different standards when owned by industrial landowners as opposed to woodlot owners, certifiers are tempted to be more lenient with industrial landowners because large landownerships represent such a monetary and prestige boost to certification.

To the extent that certification standards are vague or without measurable performance standards, this puts the power of interpretation in the hands of certifiers. This raises the question of the true independence of the certifiers. How are they chosen? Do the landowners have any influence or any veto power over who might certify them--ensuring that the certifiers are not antagonistic to industrial practices?

To what extent have the certifiers had a history of consulting for or working for companies in the forest industry? If they are consultants, to what extent is there a financial benefit to them to accept the clients' practices--because this might lead to more contracts with these or other industrial clients? To what extent have they promoted the use of controversial industrial practices such as herbicide spraying or plantations? To what extent does FSC have oversight over certifiers--who have so much power to interpret FSC standards and influence FSC credibility?

While it may not be considered polite to question the motivation or credentials of certifiers, it is hard to deny that there is a potential for conflicts and a need for some kind of oversight. Righteous indignation over such questions does not address this need.

Conclusion--rapid growth now (with hope for improvements later) or clear, measurable standards now (with growth based on a strong foundation)? I had the opportunity a number of years ago to witness a debate in the organic farming movement over certification in Maine. Very few farms in Maine were certified. One faction claimed that certification standards were too strict. They claimed that there should be an attempt to make it easier for large farms to make the transition by allowing certain chemicals at certain times. I, and some others, countered that the guidelines for chemical use suggested were too confusing, both to consumers and to farmers. I suggested that the reason few farms were certified was due to markets--there were no major market or market advantages at the time.

The next year a company manufacturing organic baby food came to Maine. Suddenly there was a big market with a financial incentive for certification. Within a short time, organic certification mushroomed. It has been growing ever since. There was no need to water down the standards. The controversy ended.

Certification for forests should be transparent and understandable to both public and landowners. The key is to make certification so compelling that there is a market premium for certified products. If there are numerous controversies swirling around certified companies the credibility will not come.

Certification should be based on a proven track record of meeting measurable outcomes, not based on promises of future performance. Creating clear standards now that meet FSC goals is a far better strategy than getting companies on board who have questionable practices now in the hope of changing them later. This strategy, enunciated in the musical *Guys and Dolls* as "Marry the man today, change his ways tomorrow," is subject to the complaint by industry that FSC keeps moving the goal posts. Women who have tried this strategy with their husbands have had very mixed results.

Appendix VII: Related Web Sites and Magazines

Web sites

Low-Impact Forestry, Maine: www.lowimpactforestry.org --Stay in touch with the latest developments, and download sample contracts, damage assessment worksheets, and logging cost calculators. You can also network with others at this site and link to all other sites mentioned in this section and more.

Community Forest Resource Center: www.forestrycenter.org --The Community Forest Resource Center was established by the Institute for Agriculture and Trade Policy to promote responsible forest management through collaboration, support, and assitance to sustainable forestry cooperatives and associations of private forest owners. The web site has extensive links and contacts.

Diseases of trees of North America: http://forestry.about.com/cs/diseases --This site has, among other things, a work by Alex Shigo, *the* expert on tree damage. Has an extensive section (with pictures) on soil life.

Ecoforestry Institute Society: www.ecoforestry.ca --The Ecoforestry Institute Society publishes the journal *Ecoforestry* and is dedicated to promoting ecologically, socially, and economically responsible forest use that maintains and restores the complexity and diversity of the forests. They are involved in certification. Their reading room has some critical articles about certification.

Forest Conservation Portal: www.forests.org--Has world-wide information on tropical and old growth forest issues, forest certification, and other forestry issues. Has extensive links as well as chat rooms on forestry topics.

Forest Stewards Guild: www.foreststewardsguild.com--The guild is a forum and support system for foresters and other resource management professionals working to ensure ecologically responsible management.

Forest Stewardship Council: www.fscus.org-- FSC sets the standards that certifiers must use.

Low-Impact Forestry, New Brunswick: www.lowimpactforestry.com --A site put together by the Conservation Council of New Brunswick and the New Brunswick Federation of Woodlot Owners. One useful feature of this site is its extensive profiles of woodlot owners in New Brunswick who are trying to lower-impacts in their management. Lots of photos.

National Communicty Forest Center Northern Forest Region: www.ncfcnfr.net--The NCFC site has, among other items, the location of demonstration forests in the region, literature on community forestry, and how to establish landowner cooperatives.

National Nework of Forest Practitioners: www.nnfp.org--The NNFP is a grass roots alliance of rural people, organizations, and businesses finding ways to interate economic development, environmental practices, and social justice.

Scientific Certification Systems: www.scs1.com-- SCS is the biggest FSC certifier in Maine at present.

Silva Forest Foundation: www.silvafor.org--SFF is a non-profit organization based in British Columbia. SFF develops and teaches the priciples of ecosystem-based planning and ecologically-responsible forest use. The site is very strong on landscape planning. SFF is involved in certification.

SmartWood: www.smartwood.org--SmartWood has the most extensive FSC certification program in the world.

Sustainable Woods Cooperative: www.sustainablewoods.com--SWC is a forest management and value-added wood processing and marketing cooperative certified by SmartWood. It is the first business of its kind in the nation, combining certified sustainable forest management by its members and certified chain of custody sales of wood products from members' forests.

Small Woodland Owners of Maine: www.swoam.com-- SWOAM's website has extensive links and information on subjects that include managment, equipment, landtrusts, easements, taxes, and Maine forest-related legislation.

Timbergreen Forestry: www.timbergreenforestry.com--Jim Birkemeir of Timbergreen is an enthusiastic promoter of sound forestry and good marketing of forest products. He is one of the driving forces behind the Sustainable Woods Cooperative in Wisconsin.

Magazines

Atlantic Forestry Review: Dvl Publishing, Inc., Box 1509 Liverpoole, N.S. BOT 1KO, ph: 902-354-5411, web: www.countrymagazines.com. AFR covers eastern Canada and northern New England. Good articles (emphasis on woodlot management and woods coops) and useful advertisements.

Independent Sawmill and Woodlot Management: P.O. Box 1149, Bangor ME 04402-1149, ph: 207-945-9469, web: www.sawmilling.com/. Good coverage of the value-added side of woodlot management as well as reviews of equipment and techniques of both management and milling.

The Northern Logger and Timber Processor: P.O. Box 69, Old Forge, NY 13420, Ph: 315-369-3078, e-mail: nela@telenet.net. Emphasis is on timber contractors with larger equipment, but covers smaller contractors and low-impact logging as well.

Northern Woodlands Magazine: P.O. Box 471, Corinth, VT 05039-0471, ph: 802-439-6292, web: wwwlnorthernwoodlands.com. Has articles of interest on forestry issues and wildlife in northern forest area of New York and New England.

About the Author

Mitch Lansky is the author of the acclaimed critique of industrial forestry, *Beyond the Beauty Strip: Saving What's Left of Our Forests*, published by Tilbury House, 1992. He is a founder of the Maine Low-Impact Forestry Project, a writer for the *Northern Forest Forum* and *Atlantic Forestry* Review, and a contributor to several books. He was a participant in the creation of the Maine Forest Biodiversity Project's book, *Biodiversity in the Forests of Maine: Guidelines for Land Management.* He has been on various government committees and task forces, including a legislative Round Table on forest labor and economic issues. Mitch, a graduate of the University of Pennsylvania, is a long-term resident of Wytopitlock, a small forest-based community in northern Maine.